KU-075-446

STABILITY, VIBRATION AND CONTROL OF SYSTEMS

Editor-in-chief: Ardéshir Guran
Co-editor: Daniel J. Inman

Dynamics with Friction
Modeling, Analysis and Experiment
Part I

SERIES ON STABILITY, VIBRATION AND CONTROL OF SYSTEMS

Series Editors: Ardéshir Guran & Daniel J. Inman

About the Series

Rapid developments in system dynamics and control, areas related to many other topics in applied mathematics, call for comprehensive presentations of current topics. This series contains textbooks, monographs, treatises, conference proceedings and a collection of thematically organized research or pedagogical articles addressing dynamical systems and control.

The material is ideal for a general scientific and engineering readership, and is also mathematically precise enough to be a useful reference for research specialists in mechanics and control, nonlinear dynamics, and in applied mathematics and physics.

Selected Forthcoming Volumes
Series A. Textbooks, Monographs and Treatises

Dynamics of Gyroscopic Systems: Flow Induced Vibration of Structures, Gyroelasticity, Oscillation of Rotors and Mechanics of High Speed Axially Moving Material Systems
  Authors: A. Guran, A. Bajaj, G. D'Eleuterio, N. Perkins and Y. Ishida

Adaptive Control of Nonlinear Systems
  Authors: R. Ghanadan, A. Guran and G. Blankenship

Series B. Conference Proceedings and Special Theme Issues

Wave Motion, Intelligent Structures and Nonlinear Mechanics
  Editors: A. Guran and D. J. Inman

Acoustic Interactions with Submerged Elastic Structures
Part I: Acoustic Scattering and Resonances
  Editors: A. Guran, J. Ripoche and F. Ziegler

Part II: Propagation, Ocean Acoustics and Scattering
  Editors: A. Guran, G. Maugin and J. Engelbrecht

Part III: Acoustic Propagation and Scattering, Wavelets and Time-Frequency Analysis
  Editors: A. Guran, A. de Hoop and D. Guicking

Part IV: Non-destructive Testing, Acoustic Wave Propagation and Scattering
  Editors: A. Guran, A. Boström, A. Gerard and G. Maze

Structronic Systems: Active Structures, Devices, and Systems
  Editors: H.-S. Tzou, A. Guran, G. Anderson, D. J. Inman and M. C. Natori

Dynamics with Friction: Modeling, Analysis, and Experiment
  Editors: A. Guran, F. Pfeiffer and K. Popp

Proceedings of the First European Conference on Structural Control
  Editors: A. Baratta and J. Rodellar

SERIES ON STABILITY, VIBRATION AND CONTROL OF SYSTEMS

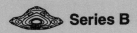 **Series B**   **Volume 7**

Series Editors: **Ardéshir Guran & Daniel J Inman**

Dynamics with Friction
Modeling, Analysis and Experiment
Part I

Editors

Ardéshir Guran
University of Southern California

Friedrich Pfeiffer
Technical University of Munich

Karl Popp
University of Hannover

 **World Scientific**
Singapore • New Jersey • London • Hong Kong

Published by

World Scientific Publishing Co Pte Ltd

P O Box 128, Farrer Road, Singapore 912805

USA office: Suite 1B, 1060 Main Street, River Edge, NJ 07661

UK office: 57 Shelton Street, Covent Garden, London WC2H 9HE

DYNAMICS WITH FRICTION: MODELING, ANALYSIS AND EXPERIMENT

ISBN 981-02-2953-4

Printed in Singapore by Uto-Print

In memoriam
Leonardo da Vinci (1452-1519)
Charles Augustine de Coulomb (1736-1806)
John William Strutt (1842-1919)

Preface

The pictures on the front cover of this book depict an example of stick-slip vibrations: the sound of a violin. Playing a violin is a complex task, involving load transmission throughout the body, intricate balance, eyes-head-neck-hand-fingers coordination and the execution of movements which leads to *stick-slip vibrations*. The quest toward understanding how we perform such tasks with skill and grace often in the presence of unpredicatable perturbations has a long history. However, despite a history of magnificent music and musical acoustics, until recent times our state of knowledge of friction induced oscillations occuring in such a basic mechanical system as well as in our everyday life was rather primitive.

The study of friction oscillators is not a mature discipline. For instance, within this book, respected leaders in the field find that understanding even simple stereotype movements in presence of friction is quite challenging. Similarly, researchers in deployable space structures find that it is surprisingly difficult to get machines to move gracefully or to interact intelligently in a dissipative environment. Yet a young violinist can perform a countless variety of tasks involving complex *non-linear dynamics of oscillations with slip-slick friction* quite effortlessly. Firction dynamics is a field that consistently humbles the researchers. Because of the complexity of friction, different friction laws are used to model different phenomena. For example, to model the self-excited oscillations in a violin bow and string, it is sufficient to express the friction as a function of sliding velocity, where the friction has a negative slope. To model stick-slip motions, the friction law used is a Coulomb law, featuring a multivalued static friction. Static friction is multivalued because it can take on a range of values to balance externally applied loads. Modern friction laws take into account variables in addition to normal load and velocity, in hope that these additional state variables can model the hysteretic effects of plastic deformation, creep and other mechanisms which may be unknown. Such friction laws have been used for analytical treatment of friction damped systems which had lead to mathematical models that are highly nonlinear, discontinuous and non-smooth. Discontinuous and/or non-smooth processes introduce problems with phase-space reconstructions, rendering analytical techniques such as perturbation methods and linearization in the vicinity of equilibrium impossible. Numerical treatments of these systems using time integration are also hampered by the non-smooth nature of the system models. Another impediment to the treatment of friction damped systems is that friction is not understood well enough to express its behavior in terms of its state variables. To this end, many researchers have made friction measurements using clever tools such as load cells, optical interference, and inverse calculations. Results have shown friction to be dependent on materials, geometry, scale, temperature, humidity, and the history of the motion (a so-called memory effect). Furthermore, friction properties have been known to show a dependence on the system in which the frictional interface is placed; this causes the frictional properties measured in test rigs to differ from those that exist *in situ*.

This book is designed to help synthesize our current knowledge regarding the role of friction in engineering systems as well as in everyday life. In some ways the task of creating this book is analogous to planning and executing a task in presence of friction We first had to define our task and its goals. We then had to work out a strategy that could help us meet these criteria. Our goals were: i) to provide contributions of high quality that span the entire field of dissipative dynamics,

ii) to serve both as a reference book for students and as a source for presentation of state-of-the-art research; iii) to synthesize as much as possible, the interrelationship between contributions; and iv) to suggest directions for future research that are likely to be fruitful.

Our strategy was manifold. First, we actively pursued leaders in the field. In this regard, we were remarkably successful. This has been clearly recognized by our contributors, and our impression has been that this situation leads to each group *putting their best foot forward*. As a state-of-the-art source book, each author was asked to include a selected review within their chapter and to keep the *Methods* section short whenever possible by referring to other publications. We also provide chapters suitable for instruction. For instance, two of us (Arde Guran and Karl Popp) will use this as a primary resource book for a graduate course entitled: *Dynamics with Friction and Damping* or as additional basic literature in a graduate course on *Nonlinear Dynamics*. To synthesize the contributions presented here, we used two approaches. First, each author had access to outlines for all other chapters. Second the book ends with an overview chapter which attempts to synthesize information within the book and to represent the most complete list of resources in friction dynamics.

In conclusion, this book provides the fruits of a team effort by leaders in this fascinating (and humbling) field of *Dynamics of Dissipative Mechanical and Structural Systems*. Put together, it provides the reader with a wealth of insight and a unique global perspective on friction dynamics. At the very least, the book shall give each of its readers a great appreciation for a task so basic as making a pleasant and enjoyable sound with a violin.

Ardéshir Guran   Friedrich Pfeiffer   Karl Popp
Washington, DC, USA   Munich, Germany   Hannover, Germany

Contributors

Jeffrey Bingham
Mechanical and Aerospace Engineering
    Department
Utah State University
Logan, Utah 84322-4130
USA

Joseph Dutson
Mechanical and Aerospace Engineering
    Department
Utah State University
Logan, Utah 84322-4130
USA

Brian Feeny
Department of Mechanical Engineering
Michigan State University
East Lansing, Michigan 48824
USA

Brook Ferney
Mechanical and Aerospace Engineering
    Department
Utah State University
Logan, Utah 84322-4130
USA

Steven Folkman
Mechanical and Aerospace Engineering
    Department
Utah State University
Logan, Utah 84322-4130
USA

Nikolaus Hinrichs
Institute of Mechanics
University of Hannover
Appelstr. 11
D 30167 Hannover
Germany

Jose Inaudi
Earthquake Engineering Research Center
University of California at Berkeley
Berkeley, California 94720
USA

James Kelly
Earthquake Engineering Research Center
University of California at Berkeley
Berkeley, California 94720
USA

Markus Oestreich
Institute of Mechanics
University of Hannover
Appelstr. 11
D 30167 Hannover
Germany

Andreas Polycarpou
Department of Mechanical and
    Aerospace Engineering
State University of New York at Buffalo
321 Gregory B. Jarvis Hall
Buffalo, NY 14260
USA

Karl Popp
Institute of Mechanics
University of Hannover
Appelstr. 11
D 30167 Hannover
Germany

Andres Soom
Department of Mechanical and
    Aerospace Engineering
State University of New York at Buffalo
321 Gregory B. Jarvis Hall
Buffalo, NY 14260
USA

Jhy-Horng Wang
Department of Power Mechanical Engineering
National Tsing Hua University
Hsin Chu, Taiwan
R.O.C.

Contents

Dynamics with Friction: Modeling, Analysis and Experiment, pp. 1–35
edited by A. Guran, F. Pfeiffer and K. Popp
Series on Stability, Vibration and Control of Systems Series B: Vol. 7
© World Scientific Publishing Company

ANALYSIS OF A SELF EXCITED FRICTION OSCILLATOR
WITH EXTERNAL EXCITATION

KARL POPP, NIKOLAUS HINRICHS AND MARKUS OESTREICH
Institute of Mechanics, University of Hannover, Appelstr. 11, D-30167 Hannover
Germany

ABSTRACT

Friction induced self-sustained oscillations, also known as stick-slip vibrations, occur
in mechanical systems as well as in everyday life. For a one degree of freedom model
of such friction oscillators the nonlinear friction force between two contacting bodies
is modelled by friction characteristics determined from experiments. The numerical
investigation of a friction oscillator with external excitation shows rich bifurcation
behaviour. Besides one- and higher-periodic response also chaotic solutions can be
observed. The influence of different types of friction characteristics is investigated.
On the basis of a one-dimensional map bifurcation and stability analysis is carried
out. The choosen way of mapping also allows a simple determination of Lyapunov
exponents. The system behaviour derived from measurements will be compared to
that of numerical investigations based on the identified friction characteristic.

1. Introduction

Self-sustained oscillations due to dry friction, also known as stick-slip vibrations, appear in everyday life as well as in engineering systems. Fig. 1 shows some
examples:

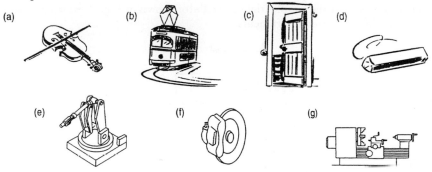

(a)   (b)   (c)   (d)

(e)   (f)   (g)

Fig. 1. Examples of stick-slip vibrations.

The sound of bowed instruments (Fig. 1(a)), the squealing noise of tramways in narrow curves (Fig. 1(b)), creaking doors (Fig. 1(c)) or squeaking chalks (Fig. 1(d)) are caused by stick-slip vibrations. Other engineering examples are rattling joints of a robot (Fig. 1(e)), grating brakes (Fig. 1(f)) or chattering machine tools (Fig. 1(g)).

The explanation of the stick-slip phenomenon can be found in textbooks, e.g. [16,21,25]. Since we are dealing with self-sustained oscillations, the vibration system encounters an energy source, an oscillator, and a switching mechanism triggered by the oscillator, which controls the energy flow from the source to the vibrating system, cf. Fig. 2.

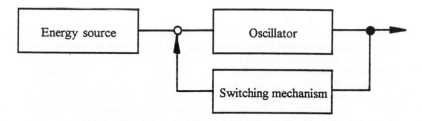

Fig. 2. Block diagram of a self-sustained vibration system.

If the energy flow into the vibration system is greater (less) than the dissipation during one period, then the vibration amplitude will increase (decrease). If the energy input and output is balanced during each period, then an isolated periodic motion occurs, which is known to be a limit cycle. Let us take the violin as an example. Obviously, the energy source is represented by the motion of the bow and the oscillator is given by the string. But how is the energy transfered from the bow to the string, or in other words, what is the switching mechanism for the energy flow? Here, the dry friction between bow and strings comes into play (this is why the bow is treated with colophony). Essential is a friction force with a decreasing characteristic for increasing relative velocities of the rubbing surfaces. This will be explained by means of the simple friction oscillator shown in Fig. 3, which has been used as a model for bowed instruments.

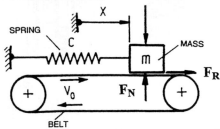

Fig. 3. Model of a friction oscillator.

The energy source is a moving belt with speed $v_0 = const$ driving the mass of the discrete spring-mass oscillator. The friction force F_R depends on the relative velocity $v_r = v_0 - \dot{x}$ between belt and mass. The friction force characteristic $F_R = F_R(v_r)$ is shown in Fig. 4. Since the limit value of the static friction force $F_R(v_r \to 0)$ is greater than the kinetic friction force $F_R(v_r \neq 0)$, the friction force characteristic is decreasing for small values of v_r. This has been observed e.g. by [5,6,12,13,22]. If the belt velocity is adjusted so that the friction force shows a negative slope for the equilibrium state $x = x_s, \dot{x} = 0$,

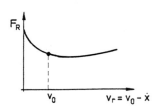

Fig. 4. Friction characteristic.

then this state becomes unstable. Thus, the amplitude grows and the trajectory in the $x, \dot{x}$-phase plane reaches fast a limit cycle, where clearly the slip mode $A \to B$ cf. Fig. 5, and the stick mode $B \to A$ can be distinguished.

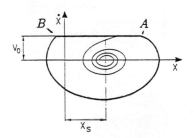

Fig. 5. Limit cycle.

The physical reason for the instability is an energy transfer from the belt to the mass. This is qualitatively shown in Fig. 6 by a sequence of time histories of one period duration. Assuming a small mass motion around the equilibrium position with a sinusoidal velocity, Fig. 6(a), results in a sinusoidal fluctuation of the relative velocity around the mean value v_0, Fig. 6(b).

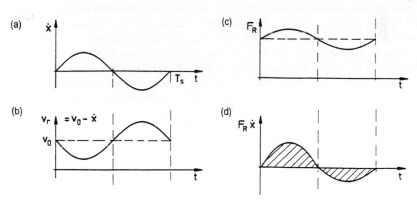

Fig. 6. Mechanical power input on the mass.

Due to the force characteristic the corresponding friction force is larger for small values v_r than for large values v_r, Fig. 6(c). Thus, the product $F_R\dot{x}$ which denotes the mechanical power input on the mass shows larger positive than negative values, Fig. 6(d). Hence, a positive energy input, $\Delta E_{in} = \int_0^{T_s} F_R \dot{x} dt > 0$, results during each period T_s, which in turn leads to increasing vibration amplitudes until the limit cycle is reached.

Fig. 7 shows three different types of friction characteristics, in the following denoted by I, II, III, and the corresponding phase portraits. Here, the friction coefficient $\mu(v_r) = |F_R(v_r)|/F_N$, i.e. the friction force F_R related to the normal force F_N in the contact area, has been plotted. For sliding, $v_r \neq 0$, the friction force is an applied force given by the friction characteristic. Whereas for stiction, $v_r = 0$, the friction force is a reaction force bounded by $|F_R(v_r = 0)| \leq \mu_0 F_N, \mu_0 = \mu(v_r = 0)_{max}$.

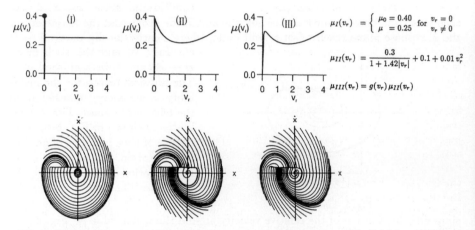

$$\mu_I(v_r) = \begin{cases} \mu_0 = 0.40 \\ \mu = 0.25 \end{cases} \text{for} \begin{array}{l} v_r = 0 \\ v_r \neq 0 \end{array}$$

$$\mu_{II}(v_r) = \frac{0.3}{1 + 1.42|v_r|} + 0.1 + 0.01 v_r^2$$

$$\mu_{III}(v_r) = g(v_r)\,\mu_{II}(v_r)$$

Fig. 7. Used friction characteristics and phase portraits with variation of the initial conditions.

Friction characteristic I is due to Coulomb's friction law and shows a static friction coefficient μ_0 and a smaller constant kinetic friction coefficient $\mu(v_r \neq 0) = \mu$. So, the transition from stick $(v_r = 0)$ to slip $(v_r \neq 0)$ is not continuous. Friction characteristic II is similar to that shown in Fig. 4 and has been explained earlier. Here, the transition from static to kinetic friction is continuous but not differentiable. Friction characteristic III follows from II by a smoothing procedure so that the characteristic is continuous and differentiable, however, there is no pronounced static friction any more and $\mu(v_r = 0) = 0$ holds. The smoothing is done by multiplication of the non-smooth characteristic with a smoothing function $g(v_r)$, where

$$g(v_r) = \frac{2}{\pi} \arctan(c_1 v_r) \tag{1}$$

has been chosen. The slope g' of the function g for $v_r = 0$ is given by

$$g' = \left.\frac{dg(v_r)}{dv_r}\right|_{v_r=0} = \frac{2}{\pi} c_1, \tag{2}$$

so the slope is proportional to c_1. The angle α between the vertical axis and the tangent to the smoothing function g in the origin is defined by

$$\alpha = \arctan \frac{1}{g'}. \tag{3}$$

The three corresponding phase portraits are qualitatively similar. Since characteristic I has no negative slope for the equilibrium state, this state does not become unstable and in its neighborhood exist periodic solutions in a pure slip mode in contrast to the solutions for characteristic II and III, where self-excitation takes place. However, for initial conditions far away from the equilibrium state, in any case all trajectories ultimately reach the limit cycle. (In case of characteristic I one better says limit curve instead of limit cycle, since other periodic solutions exist, but this distinction shall not be made in the following.) The limit cycle for characteristics I and II consists of a horizontal line due to the stick mode whereas for characteristic III there is no line representing constant velocity since there exists no real stick mode. The differences in the friction characteristics are also reflected in the transition from stick to slip and slip to stick in the corresponding limit cycles. The slope of the limit cycle in the transition points is non-smooth (smooth) for both points for characteristic I (III), whereas for characteristic II the slope is smooth for the transition from stick to slip and non-smooth for the slip to stick transition. Comparing the three phase portraits we find the interesting results, that all trajectories starting outside the limit cycle behave in a similar manner independent of the underlying friction characteristic. This robustness is important, since in applications the friction characteristic may vary considerably or is not known at all.

In engineering applications stick-slip vibrations are undesired and should be avoided since they deteriorate precision of motion and safety of operation or they are creating noise. In the present paper it is investigated how the robust limit cycle of stick-slip vibrations can be broken up by a harmonic disturbance. Thus, the self-sustained friction oscillator shown in Fig. 3 is extended by a harmonic external excitation of the base of the spring, cf. Fig. 8, and the corresponding dynamical behaviour is investigated. It is well-known that self-excited one-degree-of-freedom vibration systems with external excitation can exhibit rich bifurcation behaviour and also chaos. There exists a fast growing literature on such systems with smooth nonlinearities, e.g. [2,3,15,23,27,28,33,50]. But also in case of self-excitation due to the non-smooth nonlinearity of dry friction chaotical behaviour has been found, cf. [10,11,14,34,36,37,38,39,40,41,46,47,48]. In the following results will be shown for all three friction characteristics I, II, III. Special attention will be paid to the admissibility of smoothing procedures, which will be examined by comparing bifurcation diagrams gained for non-smooth and smoothed friction characteristics. The benefit of smoothed friction characteristics is the possibility to apply common bifurcation theory and its numerical realization in readily available computer codes, cf. [8,20,42,43]. On the other hand, for the non-smooth friction oscillator a one-dimensional map can be defined that provides an easy and illustrative method for bifurcation and stability analysis. Numerical results will be compared to those obtained from experiments.

2. Model of a self-excited friction oscillator with external excitation

The mechanical model of the friction oscillator with simultaneous self and external excitation is shown in Fig. 8.

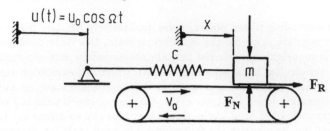

Fig. 8. Friction oscillator with self and external excitation.

The following notation is used: mass m, spring constant c, displacement of mass $x(t)$, excitation $u(t) = u_0 \cos \Omega t$, excitation frequency Ω, belt speed v_0, relative velocity $v_r = v_0 - \dot{x}$. The equation of motion reads

$$m\ddot{x}(t) + cx(t) = F_R(v_r) + cu_0 \cos \Omega t, \qquad (4)$$

where the friction force F_R depends on the relative velocity v_r and the normal force $F_N(F_N \geq 0)$. During the slip mode $(v_r \neq 0)$

$$F_R(v_r) = \mu(v_r)F_N sgn(v_r) \qquad (5)$$

holds, whereas for the stick mode $(v_r = 0)$ the friction force reads

$$F_R = c(x(t) - u_0 \cos \Omega t), \qquad (6)$$

and is bounded by

$$|F_R| \leq \mu_0 F_N. \qquad (7)$$

The static and kinetic friction coefficients follow from Fig. 7. However, for friction characteristic III no distinct stick mode occurs, thus, eq. (5) holds for the complete motion.

For the static equilibrium with $u_0 = 0$ from eq. (4) follows the static displacement

$$x = x_s = \frac{F_R(v_r = v_0)}{c} = \mu(v_r = v_0)\frac{F_N}{c}, \qquad (8)$$

cf. Fig. 5. The equation of motion can be normalized using

$$\tau = \omega_0 t, \quad \omega_0 = \sqrt{\frac{c}{m}}, \quad (*)' = \frac{d(*)}{d\tau} = \frac{\dot{(*)}}{\omega_0}, \quad v_r = v_0 - \omega_0 x'. \qquad (9)$$

From (4) and (9) it follows

$$x''(\tau) + x(\tau) = \frac{F_R(v_r)}{c} + u_0 \cos \eta\tau. \qquad (10)$$

The corresponding system of autonomous differential equations reads

$$x'_1 = x_2,$$
$$x'_2 = -x_1 + \frac{F_R(v_r)}{c} + u_0 \cos x_3,$$
$$x'_3 = \eta,$$

(11)

where $x_1 = x(\tau)$, $x_2 = x'(\tau)$, $x_3 = \eta\tau$.
For friction characteristic II and III numerical solutions have been performed on the basis of eq. (11) using the **Advanced Continuous Simulation Language, ACSL** [1]. For friction characteristics I, however, the general solution can be given analytically. For both solution methods special attention has to be paid to the exact determination of the transition states from stick to slip and from slip to stick.

3. Solution methods

3.1. Point-mapping approach

As mentioned before, for friction characteristic I an analytic solution can be formulated for the stick mode and for the slip mode. The exact states of transition from one mode to the other have to be calculated and implemented as initial values for the analytic solution of the subsequent modes. The system behaviour is calculated following the procedure shown in Fig. 9:

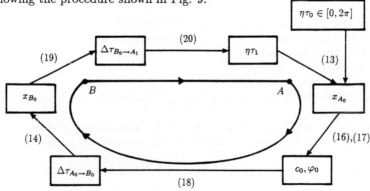

Fig. 9. Schematic diagram for the point-mapping approach.

Starting with the stick mode ($v_r = 0$, $\dot{x} = v_0$) from eq. (6) and eq. (7) it follows

$$|x(\tau) - u_0 \cos \eta\tau| \le x_0, \qquad x_0 = \mu_0 \frac{F_N}{c}.$$

(12)

The equality sign in eq. (12) holds if the spring force reaches the maximum friction force for the stick mode which characterizes the transition point A. With the choice of the excitation angle $\eta\tau_0 \in [0, 2\pi]$ from

$$|x_{A_0} - u_0 \cos \eta\tau_0| = x_0$$

(13)

the displacement x_{A_0} follows where slip initiates. Here, x_{A_i} in each cycle i is bounded by $x_0 - u_0 \leq x_{A_i} \leq x_0 + u_0$.

For the subsequent slip mode $A \to B$ $(v_r \neq 0, \mu(v_r) = \mu = const)$ to the next transition point B the corresponding solution reads

$$x(\tau) = c_0 \cos(\tau - \varphi_0) + \frac{u_0}{1 - \eta^2} \cos \eta \tau + x_s, \qquad x_s = \mu \frac{F_N}{c}, \tag{14}$$

$$v_r = v_0 - \omega_0 x' \geq 0, \tag{15}$$

where c_0 and φ_0 can be determined from the initial conditions for $\tau_0 = \tau_{A_0}$ $(x(\tau_0) = x_{A_0}, x'(\tau_0) = \frac{v_0}{\omega_0})$:

$$c_0 = \sqrt{\left(x_{A_0} - x_s - \frac{u_0}{1 - \eta^2} \cos \eta \tau_{A_0} \right)^2 + \left(-\frac{v_0}{\omega_0} - \frac{u_0 \eta}{1 - \eta^2} \sin \eta \tau_{A_0} \right)^2}, \tag{16}$$

$$\varphi_0 = -\arctan\left(\frac{\dfrac{v_0}{\omega_0} + \dfrac{u_0 \eta}{1 - \eta^2} \sin \eta \tau_{A_0}}{x_{A_0} - x_s - \dfrac{u_0}{1 - \eta^2} \cos \eta \tau_{A_0}} \right) + \tau_{A_0} \quad mod\ 2\pi. \tag{17}$$

Now eq. (14) holds till the mass velocity reaches the belt velocity again, so that the equality sign of eq. (15) is valid. Substituting the derivative of eq. (14) into eq. (15) an implicit equation for the normalized time $\Delta \tau_{A_0 \to B_0}$ of the motion from A to B is given:

$$x'(\tau_{B_0}) = -c_0 \sin(\tau_{A_0} + \Delta \tau_{A_0 \to B_0} - \varphi_0) - \frac{u_0 \eta}{1 - \eta^2} \sin(\eta \tau_{A_0} + \eta \Delta \tau_{A_0 \to B_0}) = \frac{v_0}{\omega_0}. \tag{18}$$

With the solution $\Delta \tau_{A_0 \to B_0}$ of eq. (18) the displacement x_{B_0} of point B, where the transition from the slip mode to the stick mode takes place, can be calculated from eq. (14).

Now the mass again sticks on the belt and moves with it until the restoring spring force reaches the maximum friction force acting on the mass. From eq. (12) it follows

$$|v_0 \Delta \tau_{B_0 \to A_1} - u_0 \cos(\eta(\tau_{B_0} + \Delta \tau_{B_0 \to A_1}))| = x_0 - x_{B_0}, \tag{19}$$

thus, the time $\Delta \tau_{B_0 \to A_1}$ is given implicitly. The corresponding displacement x_{A_1} of the transition point A at the time

$$\tau_1 = \tau_0 + \Delta \tau_{A_0 \to B_0} + \Delta \tau_{B_0 \to A_1} \tag{20}$$

can be calculated as mentioned above using eq. (13). Hence, the iterative procedure starts again. So, with the help of the eqs. (13) - (20) a sequence of the states x_{A_i}

and x_{B_i} can be calculated.

In contrast to a similar mapping approach used by Guckenheimer and Holmes [15] for the analysis of a bouncing ball on a vertically moving surface, here the point-mapping is given in implicit form. Thus, for solving the implicit equations a suitable Newton method has to be chosen.

3.2. Numerical integration

For friction characteristics II and III, where the kinetic friction coefficient is not constant, other solution methods have to be chosen. The system behaviour for these characteristics has been calculated by numerical integration of equation (11). For the system with characteristic II also an exact determination of the points of transition from stick to slip, A, and from slip to stick, B, is necessary. Passing these points during integration demands a special routine that determines the exact states of transition by iteration. For the smoothed friction characteristic III no distinct stick mode occurs and , thus, no transition points appear. Smoothing the friction characteristic results in stiff system equations for small values of the relative velocity. In this case appropriate integration routines and/or special stepsize control has to be used. In this work the numerical integration has been done using the program package ACSL. With the supplied language element SCHEDULE an easy and sufficient implementation of the iteration for the transition points from one mode to another is possible. By means of stepsize control algorithms the integration steps are adjusted dynamically in order to keep the error in each state variable below a prescribed value. At this point it should be mentioned that solving the system equations utilizing the point-mapping approach (if possible) leads to an enormous saving of CPU-time compared to numerical simulations.

3.3. Numerical bifurcation analysis

For smooth nonlinear systems like the friction oscillator with friction characteristic III program packages for the bifurcation analysis can be applied. Here, several computer codes are available, for instance AUTO [8], PATII [20] and BIFPACK [43]. The results of chapter 5.1 have been carried out with BIFPACK, a program for continuation, bifurcation and stability analysis.

For the task continuation and branch switching the system equation (10) must be reformulated into a boundary-value problem as follows. Periodic oscillations obey the boundary conditions

$$x(0) = x(T), \quad x'(0) = x'(T), \tag{21}$$

where T denotes the period and satisfies

$$T = n\frac{2\pi}{\eta} \quad \text{with} \quad n \in I\!N, \tag{22}$$

so that $u(\tau = 0) = u(\tau = T)$ holds.

As a consequence of eq. (22) the integration of the system equation (10) for one period requires an adapted integration interval $0 \le \tau \le T$ whenever η is changed.

Shifting the dependence of η to the right-hand side of the differential equation, the integration interval is transformed to unity length, [42]:
We introduce the normalized time ξ

$$\xi = \frac{1}{T}\tau \tag{23}$$

so that $0 \leq \xi \leq 1$ holds for one period and use the derivative

$$\frac{d(*)}{d\xi} = T\frac{d(*)}{d\tau}. \tag{24}$$

Using the notation $y_1(\xi) = x_1(\tau), y_2(\xi) = x_2(\tau)$ and $y_3(\xi) = \eta$, so that y_3 represents the bifurcation parameter, the system (11) of autonomous differential equations reads

$$
\begin{aligned}
\frac{dy_1}{d\xi} &= n2\pi\frac{y_2}{y_3}, \\
\frac{dy_2}{d\xi} &= n\frac{2\pi}{y_3}\left[-y_1 + u_0\cos(2\pi n\xi) + g(v_r)\frac{F_R(v_r)}{c}\right], \\
\frac{dy_3}{d\xi} &= 0.
\end{aligned}
\tag{25}
$$

Here, $g(v_r)$ denotes the smoothing function (1) and $v_r = v_0 - w_0 y_2$. The boundary conditions now read as follows:

$$
\begin{aligned}
y_1(0) - y_1(1) &= 0, \\
y_2(0) - y_2(1) &= 0, \\
y_3(0) &= \eta.
\end{aligned}
\tag{26}
$$

The Jacobian can be determined analytically from eq. (25). Now the required input for the use of BIFPACK is available by eqs. (25) and (26).
The procedure done by BIFPACK can be summarized as follows [42], cf. Fig. 10:

- First the initial (and final) values y_1 and y_2 of the periodic trajectories for a fixed number n have to be calculated, so that the boundary conditions are satisfied. This can be done by shooting methods:
An initial guess for $y_1(0)$, $y_2(0)$ leads, after the integration over the interval $0 \leq \xi \leq 1$, to a nonzero residual vector

$$r(y_1(0), y_2(0)) = \begin{bmatrix} y_1(0) - y_1(1) \\ y_2(0) - y_2(1) \end{bmatrix}. \tag{27}$$

Detecting the initial values $y_1(0)$, $y_2(0)$ of a periodic orbit requires a vanishing residual vector $r(y_1(0), y_2(0))$. So, a combination of an integrator, which determines the initial and final states for a set of guessed initial values, with a Newton method, that approximates the zero crossing of the residual vector calculated from the output of the integrator serves as simplest shooting method. In professional multiple shooting methods the integration interval is subdivided several times, so that better convergence of the Newton method can be obtained.

- For a suitable next step of the bifurcation parameter a predictor-method calculates an initial guess for the corresponding initial state, and the multiple shooting method leads to the determination of the periodic orbit.

- Path tracing repeats the procedure described above again and again. During continuation turning points can be approximated on the basis of a suitable test function. Additional subroutines carry out the stability analysis for calculated periodic orbits.

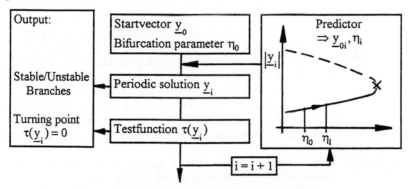

Fig. 10. Scheme for stability analysis with BIFPACK.

So, efficient methods are placed to the users disposal, that - compared to classical numerical integration - supply additional information like unstable branches and the detection of branch points.

4. System behaviour

In this section the system behaviour of the friction oscillator with simultaneous self and external excitation for the different friction characteristics I, II and III is elaborated.

4.1. Solution behaviour

The system behaviour for a fixed set of bifurcation parameters u_0, $\frac{F_N}{c}$ and η is represented by the phase trajectories. Fig. 11 shows different types of phase trajectories, where the velocity x' of the mass is shown as a function of the displacement x. The bifurcation parameters u_0 and $\frac{F_N}{c}$ have the dimension of a length. The resulting quantities x' and x have the same dimension.

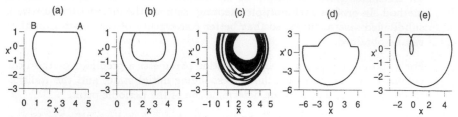

Fig. 11. Different types of phase trajectories for friction characteristic I:
(a) $\eta = 0.75$, $\frac{F_N}{c} = 10$, $u_0 = 0.5$, (b) $\eta = 2.15$, $\frac{F_N}{c} = 10$, $u_0 = 0.5$, (c) $\eta = 1.97$,
$\frac{F_N}{c} = 10$, $u_0 = 0.5$, (d) $\eta = 0.5$, $\frac{F_N}{c} = 10$, $u_0 = 4.0$, (e) $\eta = 0.41$, $\frac{F_N}{c} = 5$, $u_0 = 2.6$.

In contrast to the system without external excitation represented by the phase portraits shown in Fig. 7 the system with external excitation can exhibit one-periodic solutions (Fig. 11(a)), two-periodic solutions (Fig. 11(b)), higher-periodic solutions and also chaotic system behaviour (Fig. 11(c)) . As an interesting detail one can see that for large amplitudes u_0 of the excitation short interruptions of the stick mode become possible. Pushed by spring forces the mass overtakes the belt (Fig. 11(d)) or slows down reentering the stick mode after a small wiggle (Fig. 11(e)). In the following we confine ourselves to small excitation amplitudes u_0, so that interruptions of the stick mode as mentioned above can be excluded. For this restriction eq. (13) of the point-mapping approach can be simplified by assuming $x_{A_i} > u_0 \cos(\eta\tau)$.

The trajectory of a quasiperiodic solution in the three-dimensional state space $M = I\!\!R^2 \times S^1$ is represented by a flattened torus, cf. Fig. 12.

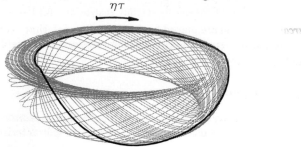

Fig. 12. Trajectory in the three-dimensional state space for friction characteristic I ($\eta = 0.6$, $\frac{F_N}{c} = 10$, $u_0 = 0.5$) and a one-periodic solution ($\eta = 0.75$, $\frac{F_N}{c} = 10$, $u_0 = 0.5$).

One can distinguish the circular ring area on top of the torus characterizing the stick mode with constant velocity of the mass ($v_r = v_0 - \omega_0 x' = 0$) and the remaining surface for the slip mode. The thick line represents the trajectory of the one periodic motion of Fig. 11(a). After one rotation around the center axis of the torus, i.e. after one period of the excitation ($n = 1$), the trajectory has rotated also one time ($p = 1$) around the body of the torus and returns to its initial state. The ratio of revolutions along and around the torus is called the winding number W, here it yields $W = \frac{n}{p} = \frac{1}{1}$.

One way to visualize the geometric structure of an attractor is the Poincaré-map, where a projection on a hyperspace Σ of the state space M is chosen that fulfills the transversality and recurrence conditions for the trajectories, e.g. [23]. For the friction oscillator the choise of Σ can be done in a simple, illustrative way: The Poincaré-map can be calculated by determining the intersection points of the trajectories on the torus M with a plane Σ transversal to the torus corresponding to $\eta\tau = const.$ In other words, the trajectory is flashed stroboscopically with the frequency of the excitation. In Fig. 13(a) the trajectory of the chaotic solution Fig. 11(c) in the three-dimensional state space is shown together with the plane Σ. From Fig. 13(b) it can be seen, that the attractor of the chaotic motion exhibits a Cantor-set-like structure, whereas the attractor for the quasiperiodic orbit from Fig. 12 is dense.

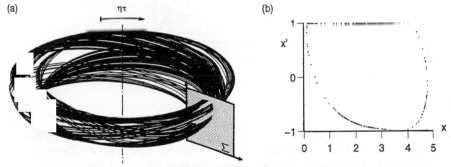

(a) (b)

Fig. 13. (a) Chaotic trajectory in the three-dimensional phase space for friction characteristic I ($\eta = 1.97$, $\frac{F_N}{c} = 10$, $u_0 = 0.5$) and (b) the corresponding Poincaré-map.

4.2. Bifurcation behaviour

For a more global examination of the bifurcation behaviour of the system, representative points of the trajectory for each set of bifurcation parameters have been extracted.

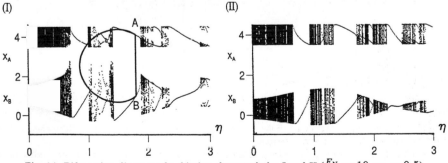

(I) (II)

Fig. 14. Bifurcation diagrams for friction characteristics I and II ($\frac{F_N}{c} = 10$, $u_0 = 0.5$).

In the bifurcation diagram the displacement x_A (transition from stick to slip) and x_B (transition from slip to stick) is plotted as a function of the bifurcation parameter η. Fig. 14(I) shows the results for friction characteristic I and Fig. 14(II) for friction characteristic II. For clarity, once again the phase plane plot of a one-periodic solution is shown in Fig. 14(I). In the bifurcation diagram of both friction characteristics one can distinguish one-periodic and higher-periodic solutions.

For large excitation amplitudes u_0 also period doubling cascades have been found, which indicate a route to chaotic motion, cf. Fig. 15.

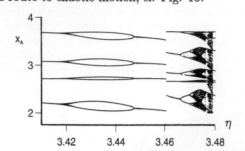

Fig. 15. Period doubling cascade for friction characteristic II ($\frac{F_N}{c} = 10, u_0 = 2.0$).

Winding number

From Fig. 12 it seems obvious that the number n of rotations of the trajectory around the center axis of the torus, i.e. the number of excitation periods, and the number p of revolutions around the body of the torus is different for different system parameters. Evaluating the winding number defined by [7]

$$W := \lim_{j \to \infty} \frac{\eta(\tau_j - \tau_0)}{2\pi j}, \tag{28}$$

leads to the interesting result shown in Fig. 16.

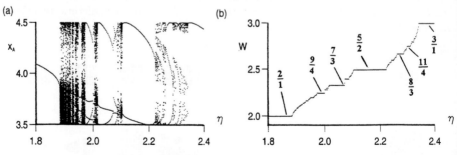

Fig. 16. (a) Bifurcation diagram for friction characteristic I ($\frac{F_N}{c} = 10, u_0 = 0.5$), (b) corresponding winding number showing a devil's staircase.

An increase of the bifurcation parameter η changes the bifurcation behaviour and results apparently in a monotone increase of the winding number. Regions of the same periodicity yield a constant winding number. In the case of a regular system behaviour the winding number is rational and given by the ratio $W = \frac{n}{p}$. Irregular system behaviour leads to an irrational winding number. The behaviour of the winding number for changing system parameters as shown in Fig. 16(b) is called a devil's staircase and has also been found in other dynamical systems, e.g. [7,35]. A diagram showing the inverse period $\frac{1}{p}$ versus the frequency ratio η is given in Fig. 17 and visualizes the self-similarity of the devil's staircase.

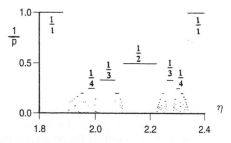

Fig. 17. Diagram showing the inverse $\frac{1}{p}$ of the oscillation period p for friction characteristic I ($\frac{F_N}{c} = 10$, $u_0 = 0.5$).

The similarity of the results for the winding number to those of the well known circle map [7,17] underlines that for the friction oscillator a one-dimensional map can be given, too.

Mapping approach

As mentioned before, for friction characteristic I an analytic solution can be formulated for the stick mode and for the slip mode. The exact states of transition from one mode to the other have to be calculated and implemented as initial values for the analytic solution of the subsequent modes. This results in a one-dimensional map H which can be explained in the following illustrative way: The limit curve of the flattened top of the torus (Fig. 18(a)) is given analytically for all values $\eta\tau$. Starting the solution procedure at any limiting point $\eta\tau_0$, after some time $\eta\tau_1$ the trajectory will return to one point $H(\eta\tau_0) = \eta\tau_1$ of the limiting curve.

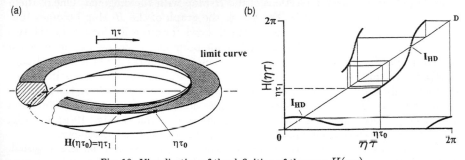

Fig. 18. Visualisation of the definition of the map $H(\eta\tau)$.

So, the map H plots the arrival point $\eta\tau_1$ as a function of the starting point $\eta\tau_0$. One can define the map H in the following way:

$$H : S^1 = [0, 2\pi] \ni \eta\tau = \eta\tau_0 \longmapsto H(\eta\tau_0) = \eta\tau_1 \in [0, 2\pi] = S^1. \tag{29}$$

Fig. 18(b) shows the graph of $H(\eta\tau)$ for the parameters $\eta = 1.55$, $\frac{F_N}{c} = 10$ and $u_0 = 0.5$. This way of description may be used to gain further insight into the different kinds of changes in the system behaviour by means of a graphical analysis. First, one has to draw the diagonal $D = \{(x, x) | x \in I\!R\}$. A vertical line from the start point $(\eta\tau_0, 0)$ of the iteration to the graph of H crosses the graph at $(\eta\tau_0, H(\eta\tau_0) = \eta\tau_1)$. Then a horizontal line from this point to D meets the diagonal at $(\eta\tau_1, \eta\tau_1)$. So, a vertical line to the graph followed by a horizontal line back to D yields the image of the point $\eta\tau_i$ under H on the diagonal. An orbit is given by drawing repeatedly line segments vertically from D to the graph and then horizontally from the graph to D. For the given set of parameters this procedure yields the following result: The orbit runs into a stable fixed point representing a one-periodic motion. The intersection points I_{HD} of the curve $H(\eta\tau)$ with the diagonal D indicate one-periodic motions. Solutions with a winding number $W = \frac{n}{p}$ will be found as intersection points of the diagonal D with $H^n(\eta\tau)$, where $H^n(\eta\tau) = H(H(H(...H(\eta\tau))))$ is the nth iterate of the map $H(\eta\tau)$.

The stability of periodic solutions can be evaluated by the slope $H'(I_{HD})$ of the H-map in the intersection point: If $|H'(I_{HD})| < 1$ then the fixed point is stable, otherwise it is unstable, cf. [24]. In contrast to the mappings known from literature, the H-map can become discontinuous and non-monotonous depending on the system parameters, cf. Fig. 18.

Intermittency

One route to chaos is intermittency. This phenomenon is characterized by intervals of apparently periodic motion interrupted by aperiodic bursts. This behaviour was first observed by [26] for the Lorenz equation, where for a critical bifurcation value η_c the stable periodic motion disappears and chaos initiates. The phenomenon intermittency can be understood if we look at some suitable parts of the H-map. Fig. 19(a) shows the situation for $\eta < \eta_c$. We can distinguish two fixed points represented by the two intersections of the H-Map with the diagonal. One of them is stable, the other is unstable. For $\eta = \eta_c$ the graph of the H-Map becomes tangential to the diagonal. In Fig. 19(b), which shows the situation for $\eta > \eta_c$, the two fixed points have disappeared in a saddle-node bifurcation. Following the iteration procedure in Fig. 19(b) near the diagonal, each iteration results in a small change of $H(\eta\tau)$. In this region the solution seems to be stable, but after some iterations the seemingly stationary behaviour changes and larger excursions from the diagonal occur. Evaluating the system response in time domain for $\eta < \eta_c$ (Fig. 19(c)) and $\eta > \eta_c$ (Fig. 19(d)) and the corresponding phase plane plots (Fig. 19(e) and Fig. 19(f)) the route to chaos via intermittency can be summerized as follows:

With increasing frequency ratio η the curve $H(\eta\tau)$ moves away from the diagonal. After a saddle node bifurcation no further stable fixed points exist and a chaotic motion starts, characterized by seemingly periodic motions interrupted by sudden bursts.

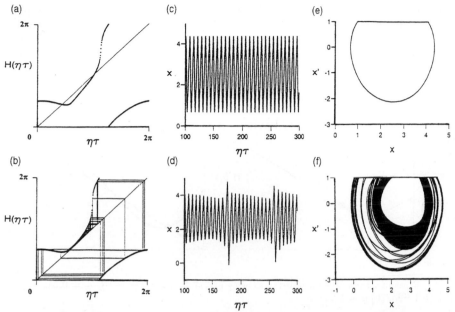

Fig. 19. Visualization of intermittency for friction characteristic I ($\frac{F_N}{c} = 10$, $u_0 = 0.5$) and $\eta = 1.8$ (upper row), $\eta = 1.9$ (lower row): (a), (b) H-maps, (c), (d) time histories of displacement x, (e), (f) phase plane plots.

Period doubling

Period doubling in nonlinear systems has been observed in all branches of physics, chemistry and biology as well as in many technical problems. Also for the friction oscillator period doubling cascades can be found (Fig. 20).

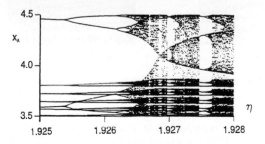

Fig. 20. Bifurcation diagram for friction characteristic I showing period doubling ($\frac{F_N}{c} = 10$, $u_0 = 0.5$).

For $\eta = 1.925$ the system exhibits a seven-periodic motion. For increasing η the system undergoes a bifurcation with a change to a periodic motion with twice the period of the original oscillation. The seven-periodic orbit becomes unstable and a stable 14-periodic orbit appears. Thereafter, a period doubling sequence occurs leading finally to chaos. One can also distinguish periodic windows. The solution in the periodic windows undergo also period doubling bifurcations, again leading to chaotic motion. The zoomed plots of the H-map show the period doubling for one of the seven stable fixed points, cf. Fig. 21(a). In the stable fixed point the slope of the H-map increases for increasing values of η. For period doubling the slope equals one. After the period doubling the former stable fixed point becomes unstable and two new stable fixed points are born, cf. 21(b).

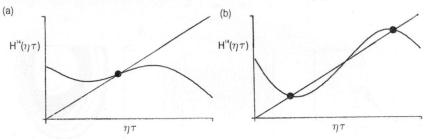

Fig. 21. H-map for friction characteristic I showing period doubling ($\frac{F_N}{c} = 10, u_0 = 0.5$), (a) $\eta = 1.925$, (b) $\eta = 1.926$.

4.3. Parameter maps

In order to give an overview of the system behaviour depending on the bifurcation parameters η and $\frac{F_N}{c}$, corresponding parameter maps have been calculated, cf. Fig. 22.

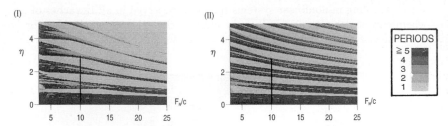

Fig. 22. Parameter maps for friction characteristic I and II ($u_0 = 0.5$).

Here, the periodicity of the solutions is visualized by a colour code. Regions of one-periodic orbits are marked in light grey, regions of five- or higher-periodic orbits including chaos are represented in black. In this diagram the results shown in Fig. 14(I) and Fig. 14(II) are visualized by solid lines. E.g. for the fixed value $\frac{F_N}{c} = 10$ on the corresponding vertical line the periodicity of the solution can be read by means of the given colour code. Thus, using these parameter maps the system behaviour can be characterized for any set of parameters within the plotted range with a fixed excitation amplitude $u_0 = 0.5$.

Comparing the results gained for the different friction characteristics shows, that the global bifurcation behaviour is independent of the kind of friction characteristic investigated. For small values of η the limit curves of the lowest two light grey regions are nearly identical. For high values of η and $\frac{F_N}{c}$ the tongues of high periodic motions disappear for friction characteristic I. The number of tongues characterizing orbits of equal periodicity is smaller for characteristic I than for characteristic II. Another interesting difference between the results for the two friction characteristics can be observed e.g. for a constant value $\frac{F_N}{c} = 10$ and $\eta > 3$. Between the regions of one-periodic and high-periodic orbits for friction characteristic I there are two-, three-, and then four-periodic solutions without higher-periodic solutions in between, in contrast to the results for friction characteristic II.

Another parameter map showing the influence of the excitation amplitude u_0 and the frequency ratio η leads to the so called Arnold tongues [17], cf. Fig. 23.

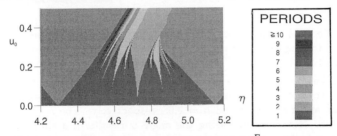

Fig. 23. Parameter map for friction characteristic I ($\frac{F_N}{c} = 10$).

For decreasing excitation amplitude u_0 the width of the tongues decreases. The frequency ratio for the tips of the tongues for vanishing excitation amplitude can be calculated by means of the winding number $W = \frac{n}{p}$.

So, by means of graphical analysis based on the point mapping approach (friction characteristic I) or numerical simulations (friction characteristic II) for each set of bifurcation parameters the corresponding solution behaviour can be determined. However, for the understanding of qualitative changes in the system dynamics by variation of the bifurcation parameters (e.g. the transition from period-one solutions to high-periodic solutions in Fig. 14 or the jump phenomenon in Fig. 15) stability and bifurcation analysis has to be carried out.

5. Bifurcation and stability analysis

5.1. Numerical analysis

The application of numerical tools for bifurcation and stability analysis like BIF-PACK [43] is restricted to smooth nonlinear systems. So, these investigations have been carried out using the smoothed friction characteristic III. In contrast to the non-smooth characteristics, for this friction characteristic a distinct stick phase does not exist. From this physical difference it cannot be taken for granted that the results gained for characteristic III using BIFPACK agree with those for characteristic II. In the following it will be investigated, whether the smoothed characteristic III is a good approximation of the original characteristic II. If the smoothing procedure is valid then the question arises: Which argument c_1 or which slope angle α should be chosen?

In Fig. 24, a comparison of bifurcation diagrams for the system with different friction characteristics is shown.

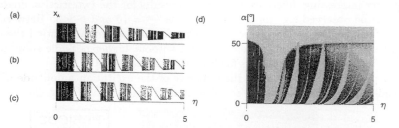

Fig. 24. Bifurcation behaviour for (a) non-smooth characteristic II, (b) smoothed characteristic III $(\alpha = 0.1^0)$,(c) smoothed characteristic III $(\alpha = 1.0^0)$ and (d) influence of the slope angle α on the periodicity of the solutions $(\frac{F_N}{c} = 10, u_0 = 0.5)$.

Fig. 24(a) belongs to the non-smooth characteristic II, see also Fig. 14(II). Fig. 24(b) and (c) show the corresponding results using the smoothed characteristic III with slope angle $\alpha = 0.1^0$ and $\alpha = 1.0^0$, respectively. The results are similar and comparable. However, increasing the angle α results in a shift of the bifurcation parameter η for periodic solutions to higher values. The bifurcation behaviour depending on η and the slope angle α is shown in the parameter map Fig. 24(d). This plot can be interpreted in the following way: For very small values of α the solution based on the smoothed friction characteristic III approximates very well the bifurcation behaviour based on friction characteristic II shown in Fig. 24(a) (compare the parameter regions for one-, two- and three-periodic orbits). With increasing slope angle α corresponding regions of periodic solutions shift to higher values of η, but the basic structure remains. For high values of α $(\alpha > 50.0^0)$ this structure is destroyed. As a simple explanation for this a look at the smoothed friction characteristic shows that in this parameter region the slope of the characteristic is positiv for all relative velocities, thus, no self excitation occurs.

With the validation of the smoothing procedure for stiff smoothing functions with $\alpha \leq 0.1^0$ further insight into the bifurcation behaviour can be obtained applying numerical bifurcation analysis. The results of the path following procedure are given in Fig. 25(a), where $y_1[u(t) = 0]$ is plotted versus η. Here, also turning points are detected and marked by crosses. The bifurcation diagram based on numerical simulations plotted in the same manner shows similar results for stable periodic solutions, cp. Fig. 25(b).

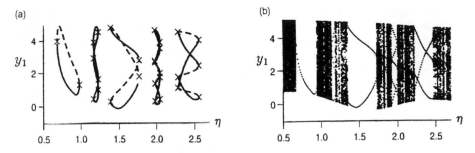

Fig. 25. Bifurcation behaviour from (a) BIFPACK, (b) simulation ($\frac{F_N}{c} = 10, u_0 = 0.5$).

However, in contrast to the bifurcation diagram Fig. 14(II) the n-periodic solutions are represented by n points because of the different type of representation chosen.

In summary, for periodic solutions there is a good agreement of the bifurcation parameters found by BIFPACK compared to numerical simulations. Additional information is given by BIFPACK with respect to the determination of unstable branches and the detection of branch points. As can be seen from the chosen way of normalisation the program package requires the knowledge of the number n. Also a periodic orbit has to be known a priori for an initial guess of the start vector for the path following algorithm.

In the following a method based on the introduced H-map is proposed that allows bifurcation and stability analysis without any a priori knowledge of the system behaviour. If $|H'(I_{HD})| < 1$ (compare chapter 4.2), the fixed point is stable, otherwise it is unstable, cf. [24]. For a periodic solution that returns after n periods of the excitation to its initial state the fixed points will be found in the iterated map H^n (and also in H^{2n}, H^{4n}, ...). The stability analysis for the non-smooth system is carried out following the procedure given in Fig. 26.

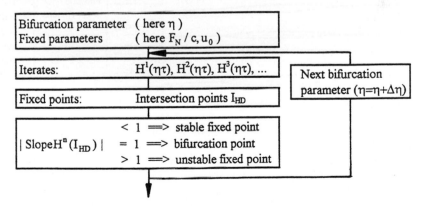

Fig. 26. Scheme for stability analysis on the basis of the H-map.

The mapping approach is not restricted to the system with friction characteristic I, where the corresponding H-map is calculated by means of the implicit equations and the slope at the intersection points follows from the implicit function theorem. For the system with friction characteristic II the corresponding H-map results from numerical integration. Here the slope at the intersection points is calculated numerically. The results of the stability analysis for friction characteristic I are given in Fig. 27(a). The comparison with Fig. 27(b) shows the same stable solutions but also the unstable solutions. The transition from one-periodic solutions to high-periodic solutions is a saddle-node bifurcation.

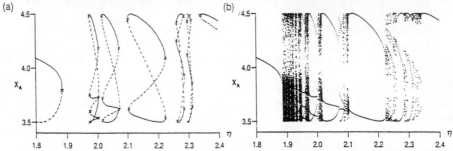

Fig. 27. Stability analysis for the non-smooth system with friction characteristic I ($\frac{F_N}{c} = 10$, $u_0 = 0.5$), (a) results from the mapping approach, (b) simulation results.

Further insight into the interesting phenomena shown in Fig. 15, i.e. jumps and period doubling, can also be gained. For friction characteristic II, in the stability diagram Fig. 28 the stable and unstable branches up to period 20 are presented. Some of the unstable branches disappear. This is due to the chosen non-smooth characteristic II which results in discontinuities of the H-map.

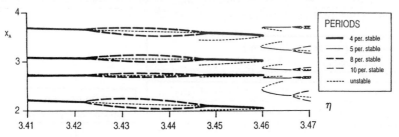

Fig. 28. Stability analysis for the non-smooth system with friction characteristic II ($\frac{F_N}{c} = 10$, $u_0 = 2.0$).

In summary, the introduced H-map supplies a simple and efficient method for bifurcation and stability analysis. For the numerical realisation of the described algorithm special attention has to be payed to the jumps of the H-map. For the existence of the jumps, cp. Fig. 18(b), two different physical explanations can be given: If the twiggle of the phase curve during sliding (Fig. 29(a)) becomes tangential to the stick line, so the mass acceleration for the transition from stick to slip equals zero, small changes of the start value $\eta\tau_0$ (Fig. 29(a)) to $\eta\tau_1$ (Fig. 29(b)) result in a discontinuity of the map (Fig. 29(c)) because of the discontinuity of the friction force.

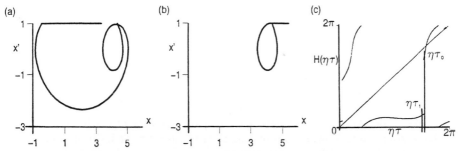

Fig. 29. First type of jumps of the H-map for friction characteristic I ($\frac{F_N}{c} = 10, u_0 = 2, \eta = 0.4$).

A look at the transition points from stick to slip gives insight into the second mechanism for the developement of the jumps. Fig. 30(a) shows the spring force for increasing x during stiction.

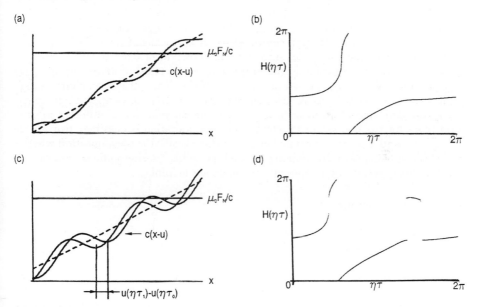

Fig. 30. Second type of jumps of the H-map for friction characteristic II ($\frac{F_N}{c} = 10, u_0 = 0.5$), (a), (b) $\eta = 1.95$, (c), (d) $\eta = 2.05$.

For small excitation amplitudes ($u_0 < \frac{v_0}{\Omega}$) the spring force curve intersects the line representing the maximum of the friction force only once independent on the angle $\eta\tau_0$. The H-map is smooth for these parameters, cp. Fig. 30(b). Shifting the start angle $\eta\tau_0$ to $\eta\tau_1$ for large excitation amplitudes $(u_0 > \frac{v_0}{\Omega})$ (Fig. 30(c)) results in a shift of the spring force curve along the line $F = cx$. This corresponds to a sudden jump of the H-map, cp. Fig. 30(d).

5.2. Comparison of methods used for bifurcation and stability analysis

BIFPACK is a useful numerical tool for bifurcation and stability analysis of smooth or smoothed nonlinear systems. The implementation of differential equations is possible in a simple way. Because of the chosen normalisation (eqs. (6) to (10)) the bifurcation analysis for the friction oscillator requires the knowlegde of the periodicity of the solution, thus, the number n must be known a priori (cp. eq. (25)). Also, a start vector near a periodic solution must be given, otherwise the algorithm fails. Another disadvantage of the path following algorithm is, that only stable and unstable solutions bifurcating from the given branch can be found. "To find all branches, good luck is also needed", [42].
Bifurcation and stability analysis for the non-smooth friction oscillator cannot be done by means of BIFPACK but on the basis of the H-map. This method does not require any a priori knowledge about the system behaviour. Another advantage of this method is, that calculating the first n iterates of the H-map, the stable and unstable branches up to period n will be found. The slope of the map for a given value $\eta\tau$ follows for friction characteristic I from the implicit function theorem and has to be calculated for friction characteristic II by means of a numerical interpolation. This can cause trouble, since for the detection of fixed points it must be decided whether the tangent to the graph $H(\eta\tau)$ is vertical or $H(\eta\tau)$ jumps. The comparison of the results gained on the basis of the H-map (friction characteristic II) and those using BIFPACK (friction characteristic III) shows a good agreement. Thus, the detection of unstable branches and the classification of different bifurcation types for the non-smooth system under consideration is possible. But for the full understanding of the dynamic behaviour there is still the open question whether the observed high-periodic motions are high-periodic, quasiperiodic or chaotic. To answer this question the Lyapunov exponents are helpful.

6. Lyapunov exponents

The determination of Lyapunov exponents is the most useful diagnostic for chaotic system behaviour. Lyapunov exponents measure the exponential rates of divergence or convergence of nearby trajectories of an attractor in the state space. In recent years methods for the calculation of the Lyapunov exponents for dynamic systems with discontinuities have been proposed in some papers, for instance [29,30]. Here, another approach is introduced. As described above, for the non-smooth three-dimensional system of the friction oscillator, the system can be reduced to a one-dimensional map H. So, by means of the system reduction we have lost two Lyapunov exponents. One of them equals zero, corresponding to a tangential perturbation direction. The other equals minus infinity, because in the stick-mode the system is two-dimensional. So, the essential Lyapunov exponent for the indication of the type of system behaviour for the non-smooth system can be determined by means of the slope of the H-map at points crossed by an orbit. An asymptotically stable orbit requires the geometric average of these slopes along the orbit to be less than one. If this average is greater than one, the orbit is unstable and therefore initially small deviations from the orbit will increase with further iterations of the map.

The measure of average local stability, the Lyapunov exponent Λ, can be given in the case of the one-dimensional map H by, cf. [18],

$$\Lambda(\eta) = \lim_{N \to \infty} \frac{1}{N} \sum_{j=0}^{N} \ln |H'(\eta \tau_j)|. \tag{30}$$

The Lyapunov exponent can be interpreted in the following way [18]:

- $\Lambda < 0$, the orbit is stable and periodic,
- $\Lambda = 0$, the orbit is neutrally stable and quasiperiodic,
- $\Lambda > 0$, the orbit is chaotic.

Fig. 31 shows the maps H for a system with friction characteristic I for a 10-periodic ($\eta = 2.0072$) and a chaotic solution ($\eta = 2.0088$) with the corresponding orbits.

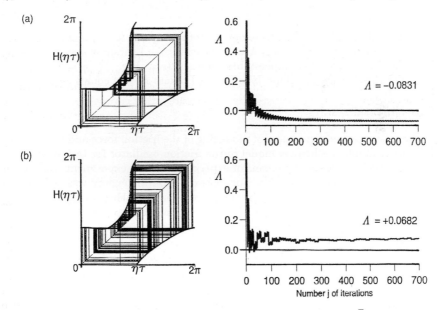

Fig. 31. Convergence of the Lyapunov exponent for friction characteristic I ($\frac{F_N}{c} = 10$, $u_0 = 0.5$), (a) $\eta = 2.0072$, (b) $\eta = 2.0088$.

The calculation of the Lyapunov exponents by eq. (30) shows good convergence with increasing number of iterations. In Fig. 32 the bifurcation diagram and the corresponding Lyapunov exponent is presented for a variation of the parameter η.

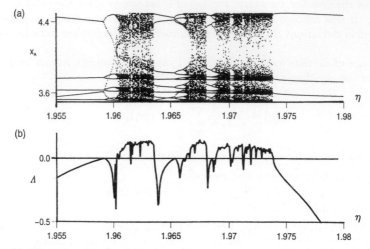

Fig. 32. (a) Bifurcation diagram, (b) corresponding Lyapunov exponent for friction characteristic I ($\frac{F_N}{c} = 10, u_0 = 0.5$).

Period doubling takes place if the slope H' of the map in a fixed point is $H' = -1$, thus the Lyapunov exponent yields $\Lambda = 0$. Whereas the Lyapunov exponent tends to $\Lambda = -\infty$ for the so called superstable solutions (e.g. Fig. 32: $\eta = 1.9601, \eta = 1.9638$). For these orbits the slope H' of crossed points $H(\eta \tau_j)$ equals zero.

7. Experimental results

In the following the correctness of modelling of the friction force and the validity of the chosen simple mechanical model of the friction oscillator for technical applications shall be investigated by means of experiments. Such experiments have been performed and reported in e.g. [38,47]. For self excitation the energy was transferred to the oscillator, a cantilever beam, by means of a moving belt similar as shown in Figs. 3 and 8. The measured signals have been processed to gain time histories, phase-plane plots, frequency spectra, the autocorrelation function, a pseudo-state space, correlation integrals and correlation dimension. However, to avoid the elasticity of the moving rubber belt a new test setup has been built, where a rotating steel disk serves as energy source.

7.1. Experimental setup

From the sketch of the experimental setup given in Fig. 33 it can be seen that the pendulum 1 with variable moment of inertia J and the friction body 2 fixed on the pendulum is pressed against the disk 3.

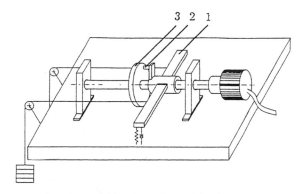

Fig. 33. Sketch of the experimental setup.

During construction of the test stand special attention has been payed to the following demands:

- possibility of change of the friction materials,
- small damping,
- variable system parameters $J, R, c_\varphi, d_\varphi, F_N, \Omega$, cf. Fig. 33.

The rotational degree of freedom of the pendulum around its center axis and the translatory degree of freedom normal to the contact plane have been realized by means of air bearings. Disk 3 can be driven with constant angular velocity Ω or periodic oscillating velocity.

The equation of motion for the rotation angle φ of the pendulum reads

$$J\ddot{\varphi} + d_\varphi\dot{\varphi} + c_\varphi\varphi = F_R(v_r)R, \qquad (31)$$

where R denotes the distance of the center of the contact area to the center axis. As can be seen from the eq. (31) the experimentally investigated system leads to the same simple mechanical model as described in section 1 with an additive (small) damping.

(a)   (b)

Fig. 34. Photograph of (a) the experimental setup, (b) the friction contact with the force transducer.

Fig. 34(a) shows photographs of the experimental setup, in Fig. 34(b) the friction contact with the three component force transducer for direct measurement of the normal and friction force is given more in detail.

Displacement and velocity of the pendulum can be measured using a laservibrometer. So, the only contact of the pendulum with solid bodies are the friction contact at the disk and the springs. The damping of the pendulum motion including material damping of the springs is very small $\left(D = \frac{d_\varphi}{2\sqrt{c_\varphi J}} < 0.001 \right)$.

7.2. Determination of friction characteristics

In order to determine friction characteristics in a first step the rotational degree of freedom has been fixed by means of another pair of air bearings. The results of friction measurements for the materials steel - brass and constant relative velocity between the contacting friction bodies is shown in Fig. 35:

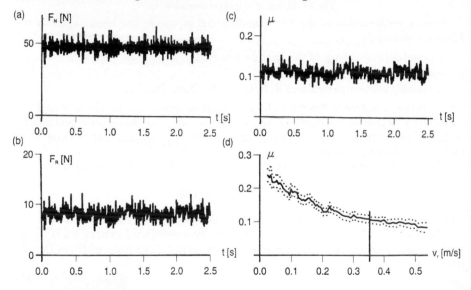

Fig. 35. Determination of the friction characteristics: (a) normal force, (b) friction force, (c) friction coefficient, (d) friction characteristic.

The normal force F_N and the friction force F_R given in Fig. 35(a) and 35(b) are oscillating which has also been found in previous works, e.g. [44]. The time dependent friction coefficient $\mu(t) = \frac{F_R(t)}{F_N(t)}$ has been calculated for each time t, the results are shown in Fig. 35(c). From the distribution $p(\mu)$ of the μ-values the mean value and the standard deviation for different values of the relative velocity v_r has been calculated and marked in the friction characteristic Fig. 35(d). The characteristic exhibits clearly a decreasing slope for increasing relative velocities. For a comparison of measured friction characteristics for different test conditions and a comparison with friction characteristics given in the literature a curve fit using an exponential function has been done. Fig. 36 shows the fitted friction characteristics for different normal loads and friction materials steel-brass (Fig. 36(a)) and steel-bronze (Fig. 36(b)).

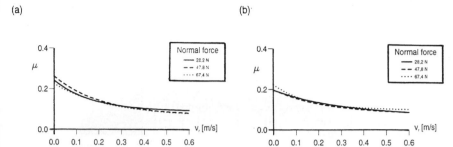

Fig. 36. Friction characteristics for different normal loads: (a) steel-brass, (b) steel-bronze.

As just known from the early observations of Amontons [4] the friction characteristics do not differ very much for different normal loads.
As described before, the disk can also be driven with harmonically oscillating velocity. The corresponding results are shown in Fig. 37.

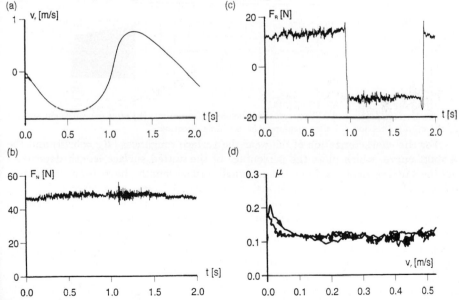

Fig. 37. Friction characteristics for oscillating relative velocity: (a) relative velocity, (b) normal force, (c) friction force, (d) friction characteristic.

Also for such test conditions the normal force oscillates (Fig. 37(b)). The friction force (Fig. 37(c)) changes its sign with a change of the sign of the relative velocity. Here, the dynamical friction characteristic follows directly from one measurement. This characteristic also shows a decreasing slope for increasing relative velocities. However, a small hysteresis occurs between measurements for increasing and decreasing relative velocity. At this point the changes in the contact surfaces due to wear should be mentioned. Measuring the contact surface before the first measurement results in the surface profile Fig. 38(a).

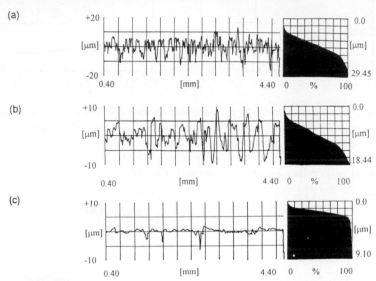

Fig. 38. Surface profile and Abbott-curve: (a) before measurement, (b) after measurement vertical to the sliding direction, (c) after measurement in sliding direction.

For the characterization of the wear the surface roughness ($R_A = 3.78$) and the Abbott-curve, which plots the percentage of the cutted surface length depending on the cutting depth z from the nominal surface length, have been calculated. After some rotations of the disk with small normal forces the surface profile and the Abbott-curve (Fig. 38(b)) vertical to the sliding direction have not changed significantly ($R_A = 3.03$). There are still grooves of about 20 μm depth. The significant change of the surface roughness can be seen in the plot of the surface roughness in sliding direction (Fig. 38(c)). Here, because of the wear the roughness has decreased ($R_A = 0.54$). The Abbott-curve shows that a larger percentage of the surface seems to support the contact partners.

In the next step we want to measure the system behaviour of the friction oscillator. The comparison of numerical simulations using identified friction characteristics with the measured system behaviour shall give a verification of the chosen model.

7.3. Measured system behaviour for self-excitation

Permitting angular motion of the pendulum, self-excitation will result in a stick-slip motion. Fig. 39(a) shows the measured phase curve for self-excitation for small and large initial conditions.

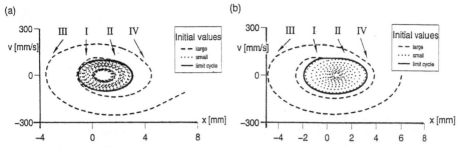

Fig. 39. Resulting system behaviour from (a) measurements, (b) simulation.

The phase curves reach fast the stable limit cycle. One can distinguish direct transitions from slip to stick (I), from stick to slip (II), transition points, where the pendulum overtakes the driving disk (III) and those, where the sign of the relative velocity changes directly (IV). The results of the numerical simulations on the basis of the identified friction characteristic is given in Fig. 39(b). A comparison of the phase curves shows a good agreement of the system behaviour found in experiments and numerical simulations.

7.4. Measured system behaviour for simultaneous self and external excitation

Additional external excitation of the self excited friction oscillator can be applied by means of an electromagnetic force generated at the end of one arm of the pendulum (Fig. 40).

Fig. 40. Photograph of the external excitation device.

Corresponding to Lorentz law a forcing moment proportional to the current in the coil moved in the magnetic field acts on the pendulum. Fig. 41 shows the time histories for two different excitation frequencies.

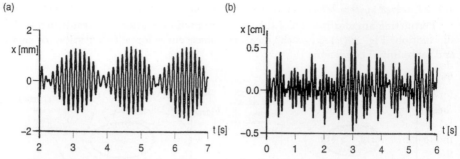

Fig. 41. System behaviour for different excitation frequencies: (a) 7.28 Hz, (b) 10.1 Hz.

In Fig. 41(a) the system response seems to be regular and the modulation of the excitation frequency and the eigenfrequency (9 Hz) can be seen. Fig. 41(b) shows a nonregular motion.

So, as seen from the simulations, changes in the system parameters result in qualitative changes of the system behaviour. Further investigations with respect to comparisons with simulation results are neccesary.

8. Conclusions

This paper deals with friction induced vibrations. Therefore, the friction characteristic plays an important role. Since this characteristic is not known exactly in reality, three different models of friction characteristics have been investigated, two of them are non-smooth. Thus, we have to cope with non-smooth dynamics. This is presently a challenge for dynamisists and mathematicians.

The friction characteristics chosen allow to build up self-sustained stick-slip vibrations, which are undesired in technical applications. It has been shown that the very robust limit cycle of stick-slip motion can be broken up by an external harmonic disturbance. The resulting friction oscillator with simultaneous self and external excitation shows rich bifurcation behaviour which has been analyzed using different methods. A first general result is, that the bifurcation diagrams and bifurcation maps for the different non-smooth friction characteristics are qualitatively similar, thus, the bifurcation phenomena are robust with respect to changes in the friction characteristics. Furthermore, the results for non-smooth friction characteristics can be approximated very well by a smoothed characteristic, if the initial slope is very steep. As a consequence, beside numerical simulation methods also computer codes for conventional bifurcation analysis can be applied. A comparison of corresponding results shows good agreement for bifurcation parameters leading to periodic motions. However, other than periodic behaviour cannot be investigated by such codes. Most promising is the third method developed and demonstrated in this paper. It is a point-mapping approach which is superior to the other methods with respect to computing time. This approach gives insight into the periodic as well as the chaotic behaviour. In summary, the method based on the developed H-map is an efficient and illustrative way to carry out the bifurcation and stability analysis for the non-smooth system under consideration. The determination of Lyapunov exponents follows without further effort. The mapping approach together with the analysis of the winding number gives deep insight into the general dynamical behaviour, although the model under consideration is based on a non-smooth mathematical description.

The presented experimental results show the non-smooth transition of the friction force for a change of the sign of the relative velocity between the contacting bodies. The identified friction characteristics exhibit a decreasing slope for increasing relative velocities. Inserting the measured friction characteristics to numerical simulations an agreement of measured and simulated system behaviour has been found.

9. Acknowledgements

The authors are grateful to Prof. Dr. A. Mielke, Institute of Applied Mathematics, University of Hannover, for his helpful discussions and valuable suggestions.

10. References

1. ACSL Advanced Continuous Simulation Language: *Reference Manual.* Edition 10.0, Mitchell & Gauthier Associates, 1991.
2. Abraham, R.H.; Shaw, C.D.: *Dynamics - the geometry of behaviour. Part 1: Periodic behavior.* Santa Cruz: Aerial Press, 1982.
3. Abraham, R.H.; Shaw, C.D.: *Dynamics - the geometry of behaviour. Part 2: Chaotic behaviour.* Santa Cruz: Aerial Press, 1983.
4. Amontons, G.: *Über den Widerstand in Maschinen.* Memoires de l' Académie Royale, pp. 203-222, 1699.
5. Bochet, B.: *Nouvelles Recherches Experimentelles sur le Frottement et Glissement.* Mines Cabur 19, pp. 27-120, 1961.
6. Conti, P.: *Sulla Resistanza die Attrito.* Accrd. Lincei. 11 (16), 1875.
7. Devaney, R.L.: *An introduction to chaotic dynamical systems.* Benjamin Cummings, California, 1986.
8. Doedel, E.J.: *AUTO: A program for the bifurcation analysis of autonomous systems.* Cong. Num. 30, pp. 265-285, 1981.
9. Feeny, B.F.; Moon, F.C.: *Bifurcation sequences of a Coulomb friction oscillator.* Nonlinear Dynamics 4, pp. 25-37, 1993.
10. Fingberg, U.: *A wheel-rail-squealing-noise model.* Proc. Polish-German Workshop on Dyn. Problems in Mech. Syst., Madralin, Poland, 1989.
11. Fingberg, U.: *A model of wheel-rail squealing noise.* J. Sound Vibr. 143, pp. 365-377, 1990.
12. Franke, J.: *Über die Abhängigkeit der gleitenden Reibung von der Geschwindigkeit.* Civil Ing., 1882.
13. Galton, I.: *The action of brakes. On the effect of brakes upon railway trains.* Engng. 25, pp. 469-472, 1878.

14. Grabec, I.: *Explanation of random vibrations in cutting on grounds of deterministic chaos.* Robotics & Computer-Integrated Manufacturing, pp. 129-134, 1988.

15. Guckenheimer, J.; Holmes, P.: *Nonlinear oscillations, dynamical systems, and bifurcation of vector fields.* New York, Berlin, Heidelberg: Springer-Verlag, 1983.

16. Hagedorn, P.: *Nichtlineare Schwingungen.* Wiesbaden: Akademische Verlagsgesellschaft, 1984.

17. Hale, J.; Kocak, H.: *Dynamics and bifurcations.* Berlin, Heidelberg, New York: Springer-Verlag, 1991.

18. Hilborn, R.C.: Chaos and nonlinear dynamics. New York, Oxford: Oxford University Press, 1994.

19. Jahnke, M.: *Nichtlineare Dynamik eines Reibschwingers.* Universität Hannover, IfM, Diplomarbeit, 1987.

20. Kaas-Petersen, C.: *PATH-User's Guide.* Leeds: University of Leeds, 1987.

21. Kauderer, H.: *Nichtlineare Mechanik.* Berlin: Springer-Verlag, 1958.

22. Kraft, K.: *Der Einfluß der Fahrgeschwindigkeit auf den Haftwert zwischen Rad und Schiene.* Archiv für Eisenbahntechnik 22, pp. 58-67, 1967.

23. Kunick, A.; Steeb, W.-H.: *Chaos in dynamischen Systemen.* Mannheim, BI, 1986.

24. Leven, R.W.; Koch, B.-P.; Pompe, B.: *Chaos in dissipativen Systemen.* Berlin: Akademie-Verlag, 1982.

25. Magnus, K.: *Schwingungen.* Stuttgart: Teubner, 1961.

26. Manneville, P.; Pomeau, Y.: *Intermittent transition to turbulence in dissipative dynamical systems.* Comm. Math. Physics 74, pp. 189-197, 1980.

27. Moon, F.C.: *Chaotic vibrations.* New York: John Wiley & Sons, 1987.

28. Moon, F.C.: *Chaotic and fractal dynamics.* New York: John Wiley & Sons, 1992.

29. Müller, A.; Hubbuch, F.: *Lyapunov Exponenten in nicht-glatten dynamischen Systemen.* To appear in ZAMM, 1995.

30. Müller, P. C.: *Calculation of Lyapunov exponents for dynamic systems with discontinuities.* To appear in Chaos, Solitons & Fractals, 1994.

31. Narayanan, S.; Jayaraman, K.: *Chaotic motion in nonlinear system with Coulomb damping.* Proc. IUTAM Symposium on Nonlinear Dynamics in Engineering Systems, Stuttgart, pp. 217-224, 1989.

32. Narayanan, S.; Jayaraman, K.: *Chaotic vibration in a nonlinear oscillator with Coulomb damping.* J. Sound Vibr. 146, pp. 17-31, 1991.

33. Neimark, J.J.; Landa, P.S.: *Stochastic and chaotic vibrations.* (In Russian), Moscow, Nauka, 1987.

34. Oestreich, M.; Hinrichs, N.; Popp, K.: *Bifurcation and stability analysis for a non-smooth friction oscillator.* Submitted to Archive of Applied Mechanics, 1995.

35. Parlitz, U.; Lauterborn, W.: *Period-doubling cascades and devil's staircases*

of the driven van der Pol oscillator. Physical Review A, Vol. 36, No. 3, pp. 1428-1434, 1987.

36. Popp, K.; Schneider, E.; Irretier, H.: *Noise generation in railway wheels due to rail-wheel contact forces.* Proc. 9th IAVSD-Symp., Linköping, pp. 448-466, 1985.

37. Popp, K.; Stelter, P.: *Nonlinear oscillations of structures induced by dry friction.* In: Proceedings of IUTAM Symposium on Nonlinear Dynamics in Engineering Systems. Stuttgart, 1989.

38. Popp, K.; Stelter, P.: *Stick-slip vibrations and chaos.* Phil. Trans. R. Soc. London A, 1990.

39. Popp, K.: *Chaotische Bewegungen beim Reibschwinger mit simultaner Selbst- und Fremderregung.* ZAMM 71, 4, pp. 71-73, 1991.

40. Popp, K.; Hinrichs, N.; Oestreich, M.: *Dynamical behaviour of a friction oscillator with simultaneous self and external excitation.* To appear in Sadhana, Bangalore, 1995.

41. Popp, K.; Hinrichs, N.; Oestreich, M.: *Numerische Untersuchung von Stick-Slip-Bewegungen mit Hilfe geglätteter Reibkennlinien.* ZAMM 75, pp. 63-64, 1995.

42. Seydel, R.: *From equilibrium to chaos; practical bifurcation and stability analysis.* Amsterdam: Elsevier, 1983.

43. Seydel, R.: *BIFPACK - A program package for continuation, bifurcation and stability analysis.* Mathematische Institute der Julius-Maximilians-Universität Würzburg, Version 2.4, 1993.

44. Soom, A.; Kim, C.-H.: *The measurement of dynamic normal and frictional contact forces during sliping.* ASME 81-DET-40, 1981.

45. Sparrow, C.: *The Lorenz equations: bifurcations, chaos, and strange attractors.* Applied Mathematical Sciences 41, New York: Springer-Verlag, 1982.

46. Stelter, P.; Popp, K.: *Chaotic behaviour of structures excited by dry friction forces.* Proc. Workshop on Rolling Noise Generation, Berlin, 1989.

47. Stelter, P.: *Nichtlineare Schwingungen reibungserregter Strukturen.* VDI R. 11 Nr. 137, Düsseldorf, 1990.

48. Stelter, P.; Sextro, W.: *Bifurcations in dynamic systems with dry friction.* Int. Ser. Num. Math. 97, pp. 343-347, 1991.

49. Stelter, P.: *Nonlinear vibrations of structures induced by dry friction.* Nonlinear Dynamics 3, pp. 329-345, 1992.

50. Thompson, J.M.T.; Stewart, H.B.: *Nonlinear dynamics and chaos.* Chichester, John Wiley & Sons, 1986.

Dynamics with Friction: Modeling, Analysis and Experiment, pp. 36–92
edited by A. Guran, F. Pfeiffer and K. Popp
Series on Stability, Vibration and Control of Systems Series B: Vol. 7
© World Scientific Publishing Company

THE NONLINEAR DYNAMICS OF OSCILLATORS
WITH STICK-SLIP FRICTION

BRIAN FEENY

Department of Mechanical Engineering, Michigan State University
East Lansing, Michigan 48824 USA

ABSTRACT

Stick-slip oscillators represent a special class of mechanical systems. They are modeled with a discontinuous velocity field, and they involve a collapsing phase space.
In single-degree-of-freedom systems, this leads to an underlying one-dimensional
map. This chapter focuses on a single-degree-of-freedom system with friction. It
discusses the relationship between stick-slip, the discontinuity, and the underlying one-dimensional maps. It connects the mechanical characteristics of a chaotic
stick-slip oscillator to the mathematical theory of one-dimensional maps and symbol dynamics. Finally, the chapter points out consequences of stick-slip on system
geometry, including implications regarding the analysis of experimental data. Most
of the discussions take place through a running example.

1. Introduction

Stick-slip is important in mechanical systems. It is typically associated with
friction. Examples of frictional systems include robot joints, braking systems, automotive squeak, rail-wheel contacts, micromachines, machine-tool processes, earthquake faults (see Chapter 3), space structures (Chapter 4), and turbine blades
(Chapter 5). Ibrahim[1] and Armstrong-Hélouvry et al.[2] have provided thorough
surveys on dynamical systems with friction.

As with Chapter 1, this chapter focuses on the dynamics of oscillators with stick-
slip dry friction. In this context, a "stick" refers to an event in which the relative
velocity between contact surfaces is zero for an interval of time. Stick-slip may also
take place in systems with lubricated contacts[3]. However, stick-slip is not confined
to friction phenomena. Many of the ideas presented here, although in the context
of friction, may be applicable to other stick-slip systems.

From the point of view of nonlinear dynamics, stick-slip has important consequences. During a stick, there is a collapse in the dimension of state space[4,5,6].
This can be visualized in state space by imagining that one of the states, velocity,
is directly constrained during a stick. For a forced one-degree-of-freedom oscillator, this produces an underlying one-dimensional map[4,6,7,8]. Such maps have been
studied extensively by mathematicians[9,10]. Thus, we have a class of mechanical
systems which corresponds to a special class of mathematical systems. The theory

developed by the mathematicians for one-dimensional maps may be applied by the engineers to the mechanical systems.

A phenomena similar to sticking can occur in impacting systems also, and is sometimes referred to as *dwell*. For example, a model of a bouncing ball will chatter, that is undergo infinitely many impacts in finite time, before resting on the table. An oscillating table will soon throw the ball into motion again[11]. Szczygielski and Schweitzer[12] observed such a dimensional collapse in an impacting rotor. In terms of the collapsing phase space, dwell is more extensive than stick-slip. During a dwell, the displacement is constrained in addition to the velocity. Thus, the collapse reduces the dimension by two.

Another related phenomenon is that of sliding-mode dynamics. While a frictional discontinuity tends to occur when relative velocities are zero, a more general system can have an arbitrarily oriented discontinuity. This is typical of sliding-mode and bang-bang control systems, in which a discontinuity is formed, by design, via a control algorithm. In the limit of a continuous-time controller, the motion can be constrained within the surface of the discontinuity, leading to sliding-mode dynamics[13]. This is an example of a dimensional collapse.

This chapter summarizes the author's participation in this subject. Section 2 discusses some examples of oscillators showing stick-slip motion. We will then focus on one of these oscillators. Section 3 examines this oscillator experimentally, and section 4 numerically. In section 5, the geometry of stick-slip is discussed with respect to this oscillator. Section 6 studies the one-dimensional map dynamics through symbol sequences and bifurcation sequences. Section 7 addresses some experimental issues.

2. Examples of Stick-Slip Oscillators

A standard example is a forced mass-spring system with Coulomb damping (Figure 1). The nondimensionalized equation of motion is

$$\ddot{x} + 2\zeta\dot{x} + x + n(x)f(\dot{x}) = a\cos\Omega t, \tag{1}$$

where x is the displacement, ζ is the damping ratio, and $a\cos\Omega t$ represents a harmonic excitation. $f(\dot{x})$ represents the coefficient of friction, and is given by

$$f(\dot{x}) = \mu\,\text{sign}(\dot{x}), \quad \dot{x} \neq 0, \qquad -1 \leq f(\dot{x}) \leq 1, \quad \dot{x} = 0. \tag{2}$$

For a constant normal load, $n(x) = 1$. Stick-slip was noticed by Eckolt[14] as early as in 1920. Periodic solutions with and without sticking were then formulated by Den Hartog[15] in 1931. More recently, Shaw[4] extended this work with a modern stability analysis, and noticed that the stick-slip dynamics reduced to that of a one-dimensional map.

Allowing the normal load to vary linearly with displacement, the normal load at the friction contact is given by

$$n(x) = 1 + kx, \quad x > -1/k, \qquad n(x) = 0, \quad x < -1/k. \tag{3}$$

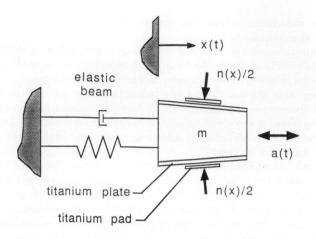

Figure 1: Mechanics model for a forced oscillator with dry friction. The friction plates are fixed to the mass m, and slide relative to the friction pads, which are fixed in x. The friction surfaces can be arranged such that they are not parallel in the direction of displacement x. In such case, the elastically loaded normal forces vary with displacement.

To prevent the existence of a negative normal load (and negative friction), the model allows for a loss of contact. This oscillator can undergo stick-slip chaos on a branched manifold, which has an underlying single-humped 1-D map[16,6]. We will look at this oscillator in great detail throughout the rest of this chapter.

The model of a belt-driven, forced oscillator also undergoes stick-slip chaos[8,17] (see Chapter 1). It has an underlying 1-D map which resembles a circle map. As with circle maps, the oscillator exhibits intermittency as a route to chaos.

Other types of systems may exhibit stick-slip. One fascinating example is a beam with a magnetic tip oscillating near a superconductor. Such a configuration produces a locus of fixed points, which provides an opportunity for sticking. A return map of this oscillator, when driven to chaos, uncovers a one-dimensional map[18]. Another example is in a beam with a plastic member[19], which also undergoes stick-slip-like behavior, and leads to reduced-order map dynamics.

The upshot is that forced single degree-of-freedom oscillators with stick-slip represent a class of systems in which the underlying one-dimensional dynamics as a typical feature. Distributed systems with stick-slip can undergo very complicated behavior, such as spatio-temporal chaos[20,21]. However, the remainder of this chapter focuses on simple friction oscillators modeled by equations (1), (2), and (3).

3. Experimental Oscillations in a Mass/Beam System

Leonardo da Vinci (1452-1519) noticed that the force of sliding friction is roughly proportional to the normal load at the contact surface. In this section, we examine an experimental system which exploits this idea.

The experiment consisted of a mass attached to the end of a cantilevered elastic beam. The mass had titanium plates on both sides, providing surfaces for sliding friction. Spring-loaded titanium pads rested against the titanium plates. We used titanium because it is relatively light and tough, and therefore a candidate for flight and space applications[22]. The coefficient of static friction for titanium on titanium is in the range of 0.7 to 1.1[23]. The titanium plates were not parallel in the direction of sliding, and thus a displacement of the mass caused a change in the force on the spring-loaded pads. Hence a change in displacement caused a change in normal load and frictional force. The angle in the titanium plates was 2.9 degrees. The elastic beam, mass and pressure pads were fixed to a common frame which was excited harmonically by an electromagnetic shaker. Strain gages attached to the elastic beam were used to sense the displacment of the mass relative to the oscillating frame. The beam and mass had a fundamental natural frequency of 2.4 Hz with the friction removed. The frequency of the second mode was 37 Hz. The excitation frequency was typically in the range of 2.5 Hz to 6 Hz, with an amplitude typically in the range of 8-12 mm. The first-mode damping ratio was $\zeta \approx 0.015$.

Further details regarding the experimental setup can be found in the literature[16].

3.1. Dynamical Behavior

Both periodic and chaotic dynamics were observed in this system. The principal periodic motions observed were of periods one and two. The experiment seemed to be inherently noisy (i.e. small high-frequency signals on the low-frequency oscillators), so period-four and higher subharmonics were not observed for a sufficiently long time to make good measurements. Speculatively, this noise may be intrinsic to friction. Contact surfaces have a random irregularity at some small scale. During the sliding process, the temperature of the contacts may vary, inducing a variation in the friction properties. Also, wear may effect the experiment by continually changing the number and distribution of contact points between the surfaces. Strain hardening and oxidation may change the amount of plastic deformation at the asperities. Some of these phenomena may contribute to hidden, unseen state variables in the modeling of friction. The "noise" may in fact consist of deterministic chaos involving such hidden variables.

A phase portrait of the oscillator, driven at 3.7 Hz, is shown in Figure 2. The motion seems to go through a funnel structure, something like with the Rössler attractor[24]. Unlike the Rössler attractor, sticking motion tends to take place within the funnel structure at positive displacements, where the normal load (and hence the friction force) is high. As the system is excited, the mass tends to creep out of this region of high friction, toward a region of lower friction, producing the funnel

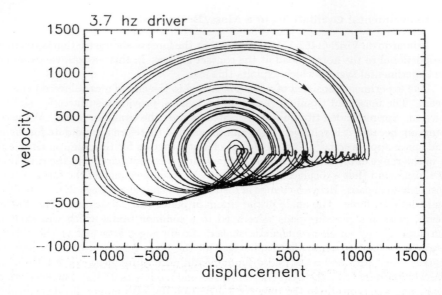

Figure 2: Phase portrait for the experimental oscillator during chaotic motion driven at 3.7 Hz, with an input amplitude of 6.3 mm.

shape. Eventually, the friction is low enough for the mass to enter into a larger orbit, which may take it back to the high-friction sticking funnel.

Sticking motion in this oscillator may be associated with heavy dissipation. In the familiar case of a block sliding freely on a surface, energy is lost during sliding, until motion stops, at which time the kinetic energy is zero. However, in the forced oscillator, the simulation model is in some sense not dissipative (by Coulomb effects) during sliding (see section 5.2). However, when the oscillator is stuck, all energy associated with the input excitation is absorbed without producing motion. Hence, there is heavy dissipation during a stick.

Quantitative statistical measurements can be found in Feeny and Moon[16].

3.1.1. Three-dimensional flow and one-dimensional map

Since the oscillator is driven periodically, we can look at the motion in $R^2 \times S^1$, where R^2 consists of displacement and velocity, and S^1 represents periodic time $t(\text{mod } 2\pi/\Omega)$, with Ω denoting the driving frequency. A projection of the three-dimensional phase portrait is shown in Figure 3. The toroidal image is similar to a deformed, flattened donut. In a cylindrical coordinate system, the time variable moves clockwise around the donut, the displacement variable goes radially away from the center, and the velocity points up.

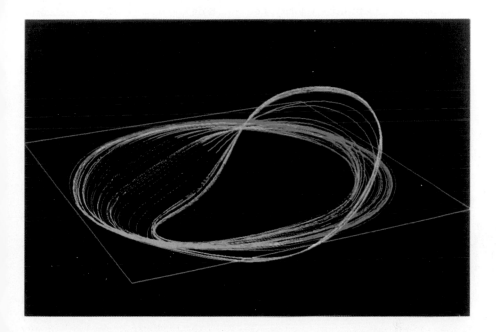

Figure 3: 3-D representation of motion for the experimental oscillator driven at 4.2 Hz. The radial, circumferential, and longitudinal axes are displacement, time, and velocity. Positive, negative and zero velocities are in white, yellow and green. Trajectories flow clockwise.

Geometrically, the motion can be approximately described by a sheet which, as time evolves, stretches and folds (Figure 4(a)), forming a branched manifold. After a driving period has evolved, the end of the sheet is folded and identified with the beginning, and it is wrapped around and attached, so that it looks like a flattened, deformed torus (Figure 4(b)). Trajectories on the sheet orbit the hole as time evolves. Motions on the outside edge of the object wind around in time, and return slightly inside the outside edge. Motions on the inside edge wind in time, push up into positive velocity, and return on the outside edge (overshooting it slightly in the experimental motion). Trajectories on the region between undergo stretching and folding.

A slice through the S^1 component of phase space at a fixed time defines a Poincaré section. It provides a cross-sectional view of the dynamical object and aids in visualizing the attractor. A Poincaré section for the motion driven at 3.7 Hz is shown in Figure 5(a). The cross section of the attractor seems to be confined to a one-dimensional object bent in two-dimensional space. It is possible to define a coordinate s along the one-dimensional Poincaré plot and to construct a delay map from the resulting sequence of points. This delay map is shown in Figure 5(b). It is a single-humped map, like the logistic map (defined by $s_{n+1} = as_n(1 - s_n)$) and the tent map (defined by $s_{n+1} = as_n$ for $0 \leq s_n \leq \frac{1}{2}$ and $s_{n+1} = a(1 - s_n)$ for $\frac{1}{2} < s_n \leq 1$). The implication is that the dynamics of a mass-beam system with displacement-dependent dry friction may be approximately reduced to a non-invertible one-dimensional map! The tent map manifests the stretching and folding of the attractor. We will return to maps in sections 5 and 6.2.

There is some ambiguity in asigning s values to points near the cusp at $s = 1$. This ambiguity causes the horizontal step-like features at $s = 0.5$ and $s = 1.5$, since such points are iterated to the area of the cusp. The true map underlying the dynamics does not necessarily have these horizontal-step features.

3.1.2. Varying attractor dimension

The Poincaré section in Figure 5(a) consists of a fuzzy part and a crisp part. The crisp part contains sticking orbits, where the velocity is zero. Qualitatively, these two portions are strikingly different. Is it possible that they have different dimensions? In the experiment, it is likely that higher modes of the beam, noise, and complications in the friction surfaces are mechanisms for disrupting this "ideal" one-dimensional Poincaré section. This causes us to wonder if higher modes or additional nonlinearities, combined with friction, could produce and example of and attractor with nonuiform dimension or topology. ("Nonuniform dimension" refers here to the localized box count in the attractor[25].) This topic is currently under investigation.

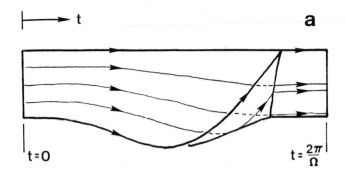

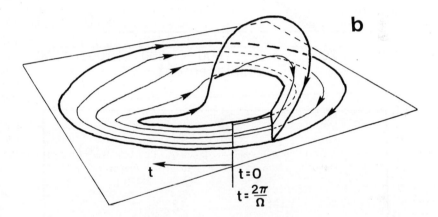

Figure 4: A geometric illustration of the motion shows trajectories confined to a sheet. (a) A sheet stretches and folds as time evolves, forming a branched manifold. (b) The beginning and end of the sheet, at time zero and after the driving period, are identified and joined to form a flattened, deformed toroidal structure. Trajectories move around the sheet, orbiting the hole. Trajectories that start on the inside fold to the outside, while those that start on the outside return near the outside.

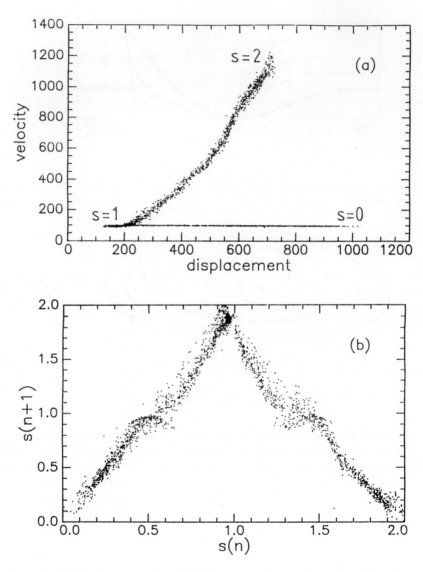

Figure 5: (a) A Poincaré section is defined as a slice in time. (b) A return map on the Poincaré section reveals a one-dimensional map.

4. Numerical Modeling

4.1. Modeling Friction

Modeling an oscillator with dry friction is not a straight-forward task. Much progress has been made in understanding the mechanisms of friction; these have been summarized in survey papers[26,1,2]. Friction descriptions range from to detailed surface tribological descriptions, to the simple Coulomb friction law, which states that there is a kinetic coefficient of friction. The former approach, although useful, is too complicated to put into a low-order differential equation of motion. On the other hand, when simple friction laws are used, dynamical friction phenomena will not all be modeled, and their effects will not be observed in the dynamics. This can effect analysis, prediction, and control[27].

The Coulomb law has been used to model stick-slip[15,4] and forced response[29]. A simple velocity-dependent friction law with a negative slope can model self-excited oscillations[30,31]. State-variable friction laws have successfully modeled hysteresis in sliding rocks[32,33,34]. Other models may be used to describe phenomena such as rising static friction[27] or normal vibrations of contact surfaces[26].

Our interest in this chapter is in stick-slip, and its effects on dynamical systems. Hence, we will focus on the Coulomb friction law, since it can model stick-slip motion (although it may not model fine details of stick-slip itself). We will also take a glance at a simple, smooth friction law, motivated by the idea of static and kinetic coefficients of friction, and a state-variable friction law which was written up based on experimental observations of the friction forces[23].

In each case, we assume that the friction force F is proportional to the normal load N, such that $F = \mu N$. In the experiment, the normal load varies with displacement, such that $N = n(x)$. In the simple friction laws the coefficient of friction is dependent on velocity, i.e. $\mu = f(v)$. Thus $F = n(x)f(v)$. If another state variable θ is involved, then the coefficient of friction might be written as $\mu = f(v, \theta)$.

With this in mind, the general nondimensional equation of motion for the mechanics model in Figure 1 is given in equation (1). The normal load is assumed to vary linearly with displacement, with the ability for the contact surfaces to separate rather than undergo tension. Thus $n(x)$ has the form of equation (3). Anderson and Ferri[28] have studied the case of bilateral contacts, with $n(x) = |1 + kx|$. For our case, normal loads in the direction of sliding have been ignored.

4.2. The Coulomb Model

People often use coefficients of static and kinetic friction μ_s and μ_k. The friction relation is given in equation (2). For simplicity, we let $\mu_s = \mu_k = \mu = 1$. Based on dynamical friction measurements in an experimental oscillator with titanium contacts, this simplification is not too unreasonable[23]. Instabilities that might occur when $\mu_s > \mu_k$ are eliminated. It turns out that such a material property is not necessary to generate chaotic dynamics. We also set the viscous damping term to zero for simplicity, and since its contribution in the experiment was small.

Before presenting numerical results, we discuss *sticking regions*, and how they effect the integration algorithm.

4.2.1. Sticking regions

At the beginning of the discussion, we will relax the condition that $n(x) = 0, x < -1/k$, and impose it again later. We first write equation (1) as a first order system. Letting $x_1 = x$ and $x_2 = \dot{x}$ yields

$$\dot{x}_1 = x_2, \qquad \dot{x}_2 = -x_1 - (1 + kx)f(x_2) + a\cos(\Omega t). \qquad (4)$$

We can determine the sticking regions by looking for fixed points. Usually, in such a non-autonomous system, there are no fixed points because $\cos(\Omega t) \neq$ constant. But here, the multi-valued friction force can balance, at least temporarily, the driving force. Looking for fixed points, we set the right-hand sides of equations (4) to zero, and obtain

$$\dot{x}_2 = 0, \qquad -x_1 - (1 + kx)f(x_2) + a\cos(\Omega t) = 0, \qquad (5)$$

whence, for $x_1 \neq -1/k$, we have

$$[x_1 - a\cos(\Omega t)]/(1 + kx) = -f(0).$$

However, $-1 \leq f(0) \leq 1$. Therefore, if

$$-1 \leq [x_1 - a\cos(\Omega t)]/(1 + kx) \leq 1,$$

the time-dependent excitation force can be instantaneously balanced by the multi-valued friction force.

For the case in which $0 \leq k < 1$, and $1 + kx > 0$ (positive normal force), we find that x_1 is temporarily fixed, i.e., x_1 is in the sticking region, when

$$x_1 \leq [1 + a\cos(\Omega t)]/(1 - k), \qquad x_1 \geq [-1 + a\cos(\Omega t)]/(1 + k).$$

For $0 \leq k < 1$, and $1 + kx < 0$ (negative normal force), a similar analysis can be carried out for the bounds on x_1.

For $k > 1$, and $1 + kx > 0$, x_1 is temporarily fixed (x_1 is in the sticking region) when

$$x_1 \geq [1 + a\cos(\Omega t)]/(1 - k), \qquad x_1 \leq [-1 + a\cos(\Omega t)]/(1 + k).$$

For $k > 1$, and $1 + kx < 0$, a similar analysis can be conducted.

In each case, the boundaries of the sticking regions are given by

$$C_1: \qquad x_1 = [-1 + a\cos(\Omega t)]/(1 + k) \qquad (6)$$

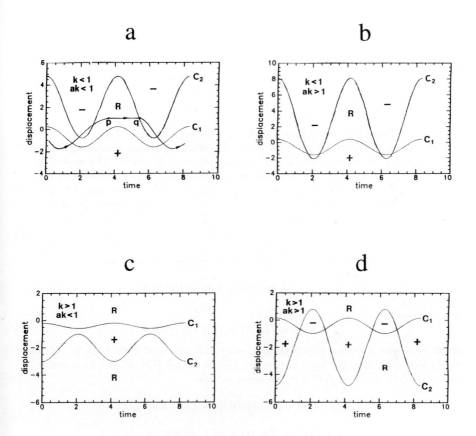

Figure 6: Sticking regions R in the (x_1, t) plane (at $x_2 = 0$) for various values of a and k: (a) $k < 1$ and $ak < 1$; (b) $k < 1$ and $ak > 1$; (c) $k > 1$ and $ak < 1$; (d) $k > 1$ and $ak > 1$. Plus and minus signs indicate where the flow is upward and downward through the page. Part (a) includes the sketch of a sticking solution, where sticking starts at point p and slip resumes at point q.

and
$$C_2: \qquad x_1 = [1 + a\cos(\Omega t)]/(1 - k), \tag{7}$$
$k \neq 1$, which intersect at $x_1 = -1/k$ when $ak \geq 1$.

The sticking regions R for various parameters a and k are displayed in Figure 6.

We now enforce $n(x) = 0, x < -1/k$ so that the surfaces can lose contact. To visualize the sticking regions in this case, we can erase all of the sticking regions plotted for $x_1 < -1/k$. For $k > 0$ there is a permanent-sticking region defined by

$$x_1 > \max(\frac{1-a}{1-k}, \frac{-1+a}{1+k}).$$

As $k \rightarrow 1^-$, the upper boundary C_1 of the sticking region in Figure 6(a) goes to $+\infty$. As k passes through the value $k = 1$, the sticking region boundary reappears from $-\infty$, and the orientation of the sticking region is inverted. The case of $k = 1$ can be specially analyzed to show that the sticking boundaries are vertical lines.

Consider an orbit which crosses the $x_2 = 0$ plane either in a sticking region or in a non-sticking zone. If it is in the sticking region, it will remain stuck, i.e., x_1 remains constant, until time evolves sugch that it is on the boundary of the sticking region. When motion resumes, will trajectories flow into $x_2 < 0$ or $x_2 > 0$? The vector field indicates the direction of flow normal to the (x, t) plane. By restricting the right side of the second of equations (4) to be greater than zero,

$$-x_1 - (1 + kx)f(x_2) + a\cos(\Omega t) > 0,$$

and we obtain

$$- x_1 + a\cos(\Omega t) > (1 + kx_1)f(0), \tag{8}$$

which describes a region in which flow *may* go from $x_2 < 0$ to $x_2 > 0$. We must satisfy this condition for both extremes, $f(0) = 1$ and $f(0) = -1$. As a result, for "upward" flow, and for $f(0) = 1$, we must satisfy

$$x_1 < [-1 + a\cos(\Omega t)]/(1 + k), \quad k \neq 1,$$

and, for $f(0) = -1$, depending on the value of k, either

$$x_1 < [1 + a\cos(\Omega t)]/(1 - k), \quad k < 1,$$

or

$$x_1 > [1 + a\cos(\Omega t)]/(1 - k), \quad k > 1.$$

To find the region of "downward" flow, i.e. flow from $x_2 > 0$ to $x_2 < 0$, we reverse the inequality in equation (8). From $f(0) = 1$, we must satisfy

$$x_1 > [-1 + a\cos(\Omega t)]/(1 + k), \quad k \neq 1,$$

and, for $f(0) = -1$, depending on the value of k, either

$$x_1 > [1 + a\cos(\Omega t)]/(1 - k), \quad k < 1,$$

or

$$x_1 < [1 + a\cos(\Omega t)]/(1 - k), \quad k > 1.$$

Thus, the curves C_1 and C_2 in the (x_1, t) plane (Figure 6) represent the boundaries of both the sticking region and the regions of upward and downward flow. This analysis will be revisited from a geometric point of view in section 5.

4.2.2. Integration algorithm

We now return to the task of describing an algorithm for numerical integration of the discontinuous equations of motion. (See Shaw[4] for a similar algorithm for friction without displacement dependence). While we discuss this algorithm particularly for this system, researchers have also been developing more general algorithms for dealing with discontinuities or unsteady topologies[35,36].

Starting with initial conditions outside of the sticking region, say $x_2 > 0$, numerical integration begins as usual with $f(x_2) = 1$, and proceeds according to the following steps.

1. Integration continues, point by point, until x_2 changes sign. The program must now remember the previous point, A.

2. A shooting routine is used, adjusting the integration step size and integrating from A to B, until B is within some tolerance of $x_2 = 0$.

3. The trajectory is now considered to be in the (x_1, t) plane; thus it is a candidate for sticking. If it is in the sticking region, the algorithm decides whether it will leave the sticking region at curve C_1 or C_2, based on the location of the point in the (x_1, t) plane. It then uses equations (6) and (7) to calculate the time of departure from the sticking region.

4. Using the current values of x_1 and t, the algorithm determines whether the onset of motion is upward $(x_2 > 0)$ or downward $(x_2 < 0)$. If upward, it sets $f = 1$ and if downward, it assigns $f = -1$. It then starts over at step 1.

Data from this scheme will not have a constant sampling rate (integration time step). In order to use data processing programs which require a constant sampling rate, the data is interpolated into data with a constant sampling rate.

The numerical integration scheme is well defined through the discontinuity, provided that the time at which a stuck trajectory begins to slip can either be calculated, or determined as infinite (implying permanent sticking). Thus, in the numerical procedure, solutions exist, and are unique in the forward sense. This idea is revisited in section 5.5.

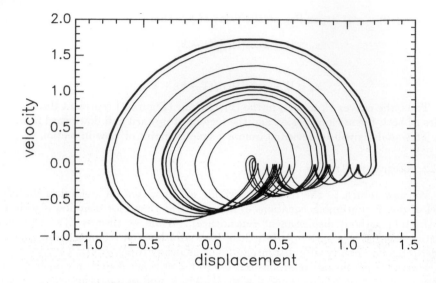

Figure 7: A numerical solution of the Coulomb oscillator with $\Omega = 1.25, a = 1.9$, and $k = 1.5$; projection on the two-dimensional $(x, \dot{x})$ space.

4.2.3. Results

The numerical study of the Coulomb model concentrates on the parameter values $k = 1.5, \Omega = 1.25$, and $a = 1.9$. The parameters are chosen not from experimentally measured values, but as values which produce behavior qualitatively similar to the experiment. Chaotic motions of this type can be seen for a large range of parameters.

The phase portrait of a chaotic numerical solution is shown in Figure 7. Sticking takes place at the trajectory cusps in the funnel structure. The system has three state variables: displacement x, velocity $\dot{x}$ and time $t(\text{mod}\, 2\pi/\Omega)$ arising from the periodic excitation. A projection of the 3-D phase portrait is shown in Figure 8. The geometry of the attractor is the same as that of Figure 4.

Taking a slice of time, we can examine the Poincaré section. Again, its image appears to be one-dimensional. We can define a coordinate s along the one-dimensional image, and plot the return map, as in Figure 9. The dynamics reduce to a one-dimensional map.

All of these features—the funnel structure, the 3-D attractor, the 1-D Poincaré image, and the tent-like return map—compare well with the experiment.

A bifurcation analysis using an increasing paramter a has shown period-doubling to be the root to chaos. Lyapunov exponents, however, were not calculated, since the presence of the discontinuity interferes with its calculation from the equations

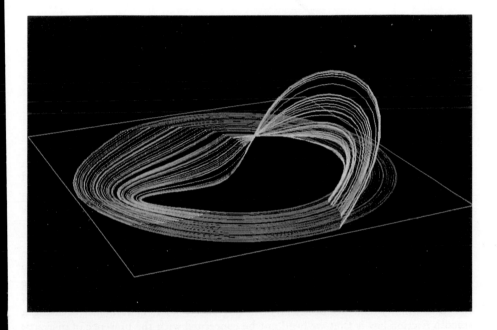

Figure 8: Numerical solution of the Coulomb friction attractor shown in 3-D with x, $t(\bmod 2\pi/\Omega)$, and $\dot{x}$ as the radial, circumferential and longitudinal coordinates. Positive, negative and zero $\dot{x}$ are in white, yellow and green. Trajectories flow clockwise.

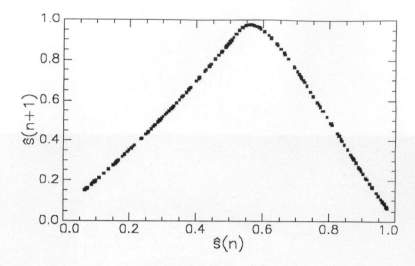

Figure 9: Return map on a coordinate s defined along the Poincaré section for a numerical simulation of the Coulomb model.

of motion. A method for computing Lyapunov exponents in the presence of a discontinuity can be found in the literature[37].

4.3. Modeling with a Smooth Friction Law

We have seen how the multivalued discontinuity led to the presence of sticking regions. It turns out that a smooth friction law with a very steep slope in place of the discontinuity can produce nearly sticking effects. The convenience of using a smooth friction law is that we need not be concerned with the presence of disconti- nuities while performing the numerical integration. To this end, we choose a smooth function for $f(v)$ which has features approximating static and kinetic coefficients of friction,

$$f(v) = (\mu_k + (1 - \mu_k)\operatorname{sech}(\beta v))\tanh(\beta v),$$

and plug this friction function into the non-dimensional equation of motion (1).

For various parameter values we can observe period-one, period-two, and higher- period motions. We can also find chaotic motions with some similarities to those of the experiment. Period doubling is the observed route to chaos. We present results for the parameter values $\Omega = 1.3, a = 1.45, k = 1.5, \zeta = 0.015, \alpha = 50, \beta = 5, \mu_k = 0.7$, and $\mu_s = 1$.

The phase portrait, displayed in 2-D, has the familiar funnel structure (Fig-

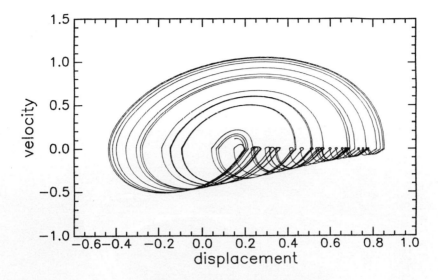

Figure 10: A numerical solution of the smooth-friction oscillator with $\Omega = 1.3, a = 1.45$, and $k = 1.5$; projection on the two-dimensional $(x, \dot{x})$ space.

ure 10). The 3-D phase flow shows a stretching and folding toroidal (Figure 11). A return map on the Poincaré section reveals a tent-like one-dimensional map (Figure 12). These features resemble those of the experiment and the Coulomb model.

True stick-slip cannot take place with the smooth friction model. Since there is no discontinuity, there are no true sticking regions. Instead of a discontinuity, there is a very steep slope at the origin of the friction function. Its effect is to strongly dampen motions in that region. As a result, "nearly sticking" motion seems to take place. This is visible at the tiny loops in the funnel of Figure 10, as compared to the cusps of Figure 7, which represent true stick. It is also visible in Figure 11 as the circular orbits near the zero-velocity plane, underneath the fold.

With the smooth function we can easily calculate a Lyapunov exponents. Since the smooth law is differentiable, the differential equation of motion is differentiable. Hence, the variational equations, which require differentiation of the vector field, can be computed. The largest Lyapunov exponent was calculated as $\lambda = 0.11$, which is greater than zero, indicating sensitivity to initial conditions and a loss of longterm predictability[38].

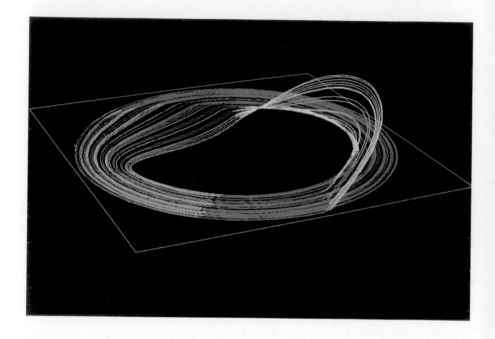

Figure 11: Numerical solution of the smooth-friction attractor shown in 3-D with x, $t(\mathrm{mod}\,2\pi/\Omega)$ and $\dot{x}$ as the radial, circumferential, and longitudinal coordinates. Positive, negative, and (nearly) zero $\dot{x}$ are in white, yellow, and green. Trajectories flow clockwise.

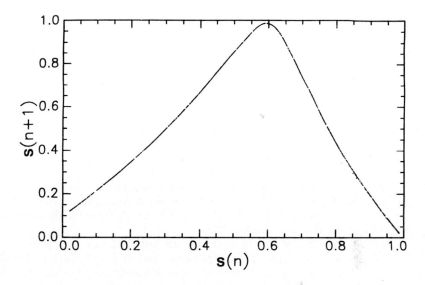

Figure 12: Return map on a coordinate s defined along the Poincaré section for a numerical simulation of the smooth-friction model.

4.4. Modeling with a State-Variable Friction Law

In the previous section we described friction laws which depend only on velocity and displacement (via the normal load). In this section, we explore a friction law which depends on an additional "hidden" variable. Such friction laws have been dubbed *state-variable friction laws*, and have been used to model experimental data on the steady sliding of rocks[32,33,34].

Why introduce an extra state variable here? Notice that in the 3-D display of the experimental dynamics (Figure 3), orbits that start on the inside of the deformed torus travel through positive velocity, and fold onto the outside of the torus, *passing through* the zero-velocity plane before returning to zero velocity. On the other hand, in the previous simulations, trajectories that start on the inside of the deformed torus travel through positive velocity, and fold onto the outside of the torus, *sealing onto* the zero-velocity plane (sticking motion). The discrepancy could be due to experimental factors, such as the effect of higher modes in the beam, or filtering. Conversely, there might be some shortcomings in the simple friction laws. In what follows, the state-variable friction model will produce the feature of overshooting the zero-velocity plane prior to achieving sticking motion.

We choose a friction model based on experimental observations of friction measurements, without regard to the physics of friction[23]. In this formulation, the equations of motion have an additional state θ, such that

$$\ddot{x} + x + n(x)\theta = a\cos(\Omega t), \qquad \dot{\theta} = -\gamma(\theta - f(\dot{x})), \tag{9}$$

where $f(\dot{x})$ is a simple friction function, such as the Coulomb law. θ represents the instantaneous coefficient of friction, and for a constant $\dot{x}$ it asymptotically approaches a backbone function, represented by $f(\dot{x})$. Thus, the friction law is like a simple one, but with some inertia.

Equations (9) use a state-variable friction law similar to those of Ruina[33,34]. The backbone is slightly different, and it allows for relative sliding to switch directions. (Ruina has proposed a modification to his law to accomodate sliding reversals). Also the right-hand side of Ruina's version of the second of equations (9) is multiplied by the velocity. Experiments with other frictional systems indicate that this type of velocity dependence is necessary to model certain friction phenomena[34]. We are also assuming that the normal load $n(x)$ has no direct effect on the dynamics of θ, contrary to the observations of others[39,40].

Equilibrium solutions of equations (9) for the unforced case with $n(x) = 1$ are $\bar{x} = -\bar{\theta}$, and $\bar{\theta} = f(0)$. Thus, if $f(\dot{x})$ is multivalued at $\dot{x} = 0$, such as with the Coulomb law, the undriven oscillator will have infinitely many equilibria. Still, during oscillation, the friction force will not change discontinuously in time.

A simulation of the oscillator with skewed plates was performed by letting $n(x) = 1 + kx, x > -1/k$ and $n(x) = 0, x < -1/k$. For simplicity, we used $f(\dot{x}) = \tanh(50\dot{x})$ (a smooth approximation of the Coulomb law with $\mu_k = \mu_s = 1$). A 3-D phase portrait is shown in Figure 13 for the parameter values of $k = 1.5, \Omega = 1.25, a = 1.9$, and $\gamma = 10$. Note, however, that equations (9) represent a four-dimensional system. In the plot, motions starting from the inside of the torus flow through positive velocity and overshoot the zero-velocity plane before achieving "nearly" sticking motion, similar to the experimental oscillator.

The state-variable friction law in equations (9) is not guaranteed to be passive. The friction will do positive work on the system whenever θ and $\dot{x}$ differ in sign. No analysis has been performed in this regard. For the large motions simulated, this does not present any problems. However, for small motions, it could be that the model and experiment will have qualitatively different behaviors.

4.5. Summary

Due to the multivalued discontinuity, the Coulomb law incorporates stick-slip in its behavior, while the smooth, simple friction law, and state-variable law approximate stick-slip. Each friction model is able to recreate qualitative features seen in the experimental system. These features include the strange attractor on a branched manifold, and the reduction of the dynamics to a one-dimensional map. The state-variable friction law was able to model the addition detail of trajectories overshooting the sticking region before approaching it. The experiment displayed this overshoot, while the simple friction models did not incorporate it.

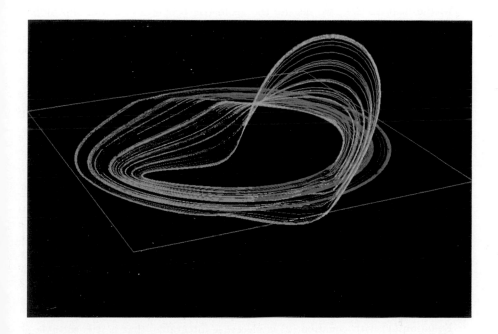

Figure 13: Numerical solution revealing the attractor for the state-variable-friction model shown in 3-D with x, $t(\mathrm{mod}\, 2\pi/\Omega)$ and $\dot{x}$ as the radial, circumferential, and longitudinal coordinates. Positive, negative, and (nearly) zero $\dot{x}$ are in white, yellow, and green. The flow is clockwise.

The Coulomb model and the smooth model suggest that the important contributors to the formation of the attractor on a branched manifold are the (near) discontinuity of the vector field, and the dependence of the fricton on displacement as well as velocity. A negative slope is not necessary for obtaining chaotic motion in a friction oscillator. Such negative rate dependence is likely to be a material friction property, while the displacement dependence can easily be a property of the mechanical set-up.

In the next section, we discuss the geometric properties of stick-slip in this oscillator, and how the one-dimensional map arises.

5. The Geometry of Stick-Slip

In this section, we study the geometry of the vector field corresponding to equations (1), (2), and (3). We have already noted that an interesting feature of this ordinary differential equation of motion is that, because of the friction function, it is *discontinuous* and *multivalued* at $\dot{x} = 0$. It is multivalued in that $f(0)$ can take on any value between $-\mu_s$ and μ_s.

One approach to such a problem is to view it as a piecewise continuous system, examine the continuous pieces, and match them. Alternatively, we might look at the limiting behavior of continuous systems which in some sense converge to the discontinuous system[41]. This section focuses on the former approach.

The case of $k = 0$, and hence $n(x) \equiv 1$, can be solved for periodic motion[15] and their stabilities[4] by breaking the problem into piecewise linear equations and exploiting symmetry in x. Unfortunately, by including $k \neq 0$, thereby causing $n(x)$ to be active and nonconstant in equation (1), we lose the symmetry in x. Hence, the calculation of periodic orbits and their stabilities is extremely difficult. We choose another approach for the analysis: we graphically examine the qualitative nature of the system. This type of analysis has been done in classic nonlinear-vibrations texts[42,30] for two-dimensional autonomous systems.

Equation (1) with equation (2) is piecewise integrable, that is it is solvable in subregions of the state space. We geometrically observe the nature of the flow in each of these regions of solvability, and then see how these solutions interact at the boundary of the regions. The dynamics of the flow are viewed in terms of a map on the boundary between the regions. From a qualitative picture of this map, we can construct the attractor, and show that

1. the dynamical behavior reduces to a one-dimensional map,

2. the flow of the Coulomb oscillator may not be invertible (previously reported by Shaw[4]),

3. the flow may reach its attractor in finite time, and

4. the attractor has dimension less than or equal to two.

Takens[43] had observed these properties in *constrained systems*[44]. These properties are made possible because the Coulomb friction law produces a discontinuous and multivalued vector field (the same mechanism responsible for stick-slip motion). Finite attraction time to "terminal attractors" has been exploited for the learning process in neural networks with non-Lipschitz components[45]. Other properties may arise in systems with discontinuities. For example, in a problem modeling shock, Antman[46] uncovered nonunique forward solutions. We will also look at the ramifications of stick-slip on experimental methods in section 7.

5.1. *Piecewise Linear Equations*

Again, we look at the special case of $\mu_s = \mu_k = 1$. Further, we neglect the viscous damping term $\zeta\dot{x}$ (recall that the experimental damping ratio was measured as $\zeta = 0.015$). Finally, out of interest, we will only look at the case where $k > 1$, for which the sticking region has a particular structure (see section 4.2.1). Equation (1) with equations (2) and (3) can be written for regions in which they are solvable:

$$\ddot{x} + (1+k)x = -1 + a\cos(\Omega t), \qquad \dot{x} > 0, x > -\frac{1}{k}, \tag{10}$$

$$\ddot{x} + (1-k)x = 1 + a\cos(\Omega t), \qquad \dot{x} < 0, x > -\frac{1}{k}. \tag{11}$$

For $x \leq -1/k$, the normal load becomes zero due to the no-contact condition, and we have

$$\ddot{x} + x = a\cos(\Omega t), \qquad x < -\frac{1}{k}. \tag{12}$$

For $-1 < k < 1$, equation (10) has harmonic solutions. For $k > 1$, the solution of equation (11) consists of a driven saddle.

Let D denote the (x,t)-plane $(\dot{x} = 0)$ and consider the mappings associated with the flow of points based at D :

$$P^+ : \oplus \to D, \quad \dot{x} > 0,$$

$$P^- : \ominus \to D, \quad \dot{x} < 0,$$

where P^+ is the map that arises from the flow of equation (10), and P^- is the map that arises from the flow of equation (11), and $\oplus \subset D$ and $\ominus \subset D$ denote the domains of P^+ and P^-, respectively. The goal is to geometrically describe the maps P^+ and P^-, and see how they interact with the sticking region. The idea is sketched in Figure 14. For simplicity, we will start the discussion by omitting equation (12), which arises from the no-contact condition. Thus we will discuss the mapping between the flow of equation (10) and equation (11). This analysis is of interest anyway, since regular and chaotic motions confined to the region $x > -1/k$ have been observed in some numerical integrations. The results will be completed by including equation (12) numerically.

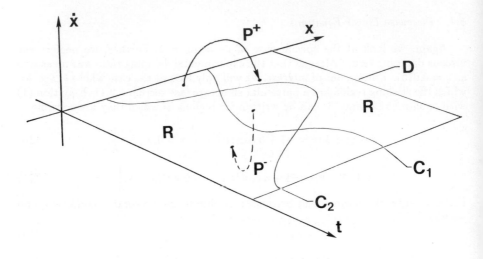

Figure 14: Trajectories governed by each piecewise linear equation are associated with either the map $P^+ : \oplus \to D$, for trajectories in $\dot{x} > 0$, or the map $P^- : \ominus \to D$, for trajectories in $\dot{x} < 0$. Some orbits get mapped into the sticking region R, where the motion remains constant until such time that the orbit is no longer in the sticking region. The curves C_1 and C_2 represent the boundaries, in D, between the sticking regions R and the domains $\oplus$ and $\ominus$ of P^+ and P^-, respectively.

5.1.1. The sticking region—geometric interpretation

Analysis of equation (1) with equation (2) for $\dot{x} = 0$ leads to the sticking regions. In section 4.2, the sticking region was found by analyzing equations (4) for fixed points and using the multivaluedness of $f(x_2)$ at $x_2 = 0$[15,4,16]. However, we will describe the sticking regions using a geometric viewpoint which will help set the mood of this analysis.

If we write equations (4) in extended phase space for $\dot{x} > 0$, we have

$$\dot{x}_1 = x_2$$

$$\dot{x}_2 = -(1+k)x_1 - 1 + a\cos(\Omega t), \quad x_2 > 0 \tag{13}$$

$$\dot{t} = 1.$$

By looking at the sign of $\dot{x}_2$ adjacent to the (x, t)-plane D, we find regions where the flow of equations (13) is upward and regions where the flow is downward. The curve (in the plane D) dividing the regions, called C_1, is given by $\dot{x}_2 = 0$, which yields

$$C_1: \quad x_1 = \frac{-1 + a\cos(\Omega t)}{1+k}.$$

We can do the same for the case of $\dot{x} < 0$. The equations in extended phase space are

$$\dot{x}_1 = x_2$$

$$\dot{x}_2 = -(1-k)x_1 + 1 + a\cos(\Omega t), \quad x_2 < 0 \tag{14}$$

$$\dot{t} = 1.$$

The regions for upward and downward flow on the (x, t)-plane D for equations (14) are separated by a curve C_2, which is given by $\dot{x}_2 = 0$, or

$$C_2: \quad x_1 = \frac{1 + a\cos(\Omega t)}{1-k}.$$

Since the flows of equations (13) and (14) meet at D, we try to match the flow above D with the flow below D. There are some regions where the flow of equations (13) is directed from $\dot{x} > 0$ toward D, and simultaneously the flow of equations (14) is directed from $\dot{x} < 0$ toward D, producing a conflict in the flow directions. (Similarly, there are regions where th flow from both $\dot{x} > 0$ and $\dot{x} < 0$ are directed away from D.) These regions of conflict are the sticking regions R, as shown in Figure 15. Regions where the flow directions agree and are upward (toward $\dot{x} > 0$) are labeled $\oplus$, and regions where the flow directions are downward (toward $\dot{x} < 0$) are labeled $\ominus$. In the region "above" (defined as $\{(x, t) : x > x^{(1)}$ and $x > x^{(2)}$, where $(x^{(1)}, t) \in C_1$, and $(x^{(2)}, t) \in C_2\}$) both C_1 and C_2, the flows of *both* equations (13) and (14) are directed *toward* the (x, t)-plane. Hence, orbits hitting this region are trapped in D, x remaining fixed, as time evolves until the moment that the direction of flows

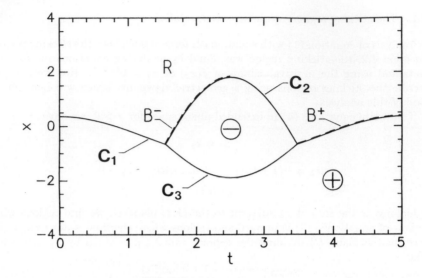

Figure 15: The sticking regions when the no-contact condition is observed. $\oplus$ and $\ominus$ indicate upward and downward flows. R indicates the sticking region. B^+ labels the portion of C_1 with positive slope, bordering $\oplus$, and B^- labels the portion of C_2 with positive slope, bordering $\ominus$. $\Omega = 1.25, a = 1.9$, and $k = 1.5$.

is in agreement. The multivaluedness of $f(x_2)$ provides a bridge between the flow of equations (13) and the flow of equations (14), and allows this trapping to take place. In the region "below" both C_1 and C_2, $x_1 \leq -1/k$, and due to the no-contact condition, there are no sticking regions.

If we apply the no-contact condition, the direction of flow for $x < -1/k$ is then governed by equation (12). The curve C_3 separating regions of upward and downward flow in equation (12) is given by

$$C_3 : x_1 = a \cos(\Omega t), \quad x_1 < -\frac{1}{k}.$$

The curves C_1, C_2, and C_3 all intersect at the same points:

$$x_1 = -\frac{1}{k},$$

$$t = \frac{1}{\Omega} \arccos(-\frac{1}{ak}).$$

5.2. Qualitative Mapping of Regions

We would like to see how the active regions (nonsticking regions) map under P^+ and P^-, and how they interact with the sticking regions[6]. Again, results are presented for parameter values of $a = 1.9, \Omega = 1.25$, and $k = 1.5$.

Referring to Figure 15, we consider the mapping of the region $\oplus$ via P^+, and the mapping of the region $\ominus$ via P^-. If R is the sticking region, then $\oplus \cup \ominus \cup R = D$. Motions in R either stay in R forever, or, through the evolution of time, exit R into $\oplus$ or $\ominus$ via the map $S : R \rightarrow B^+ \cup B^-$, where B^+ and B^- are part of the boundaries of $\oplus$ and $\ominus$ as shown in Figure 15. (The sets B^+ and B^- are the components of C_1 and C_2, respectively, through which sticking orbits originating in R must pass as they exit R.) Therefore, the system can be understood through the mappings of $\oplus$ and $\ominus$. Certainly $P^+(\oplus) \cap \oplus = \emptyset$, and $P^-(\ominus) \cap \ominus = \emptyset$. It is also likely that $P^+(\oplus) \cap R \neq \emptyset$, and $P^-(\ominus) \cap R \neq \emptyset$.

We want to qualitatively describe the images $P^+(\oplus)$ and $P^-(\ominus)$.

The dynamics of the Coulomb oscillator can be described by successive mappings of $\oplus$ or $\ominus$, under P^+, P^-, and S, when appropriate. Figure 16 shows the computer-generated sequence of mappings of $\ominus$ for a particular set of parameter values. The process accounts for no-contact condition. Within one period of time, the entire region of initial conditions has been crushed into a set of curves. This smashing is *not asymptotic!* It occurs suddenly in the sticking regions. The attractor lies in the images of B^+ and B^-. Many initial points will be condensed onto the attractor in finite time.

A three-dimensional blob of initial conditions condenses into a two-dimensional blob as it flows into the sticking region. Thus volumes of phase space are collapsed during a stick. This dimensional collapse corresponds to the heavy energy dissipation discussed in section 3.

The result in Figure 16 suggests that we only need to understand mappings of B^+ and B^- to understand the long-term behavior of the entire system. When we study the system in this way, we are implicitly assuming that all orbits eventually pass through the sticking region, exiting onto B^+ and B^-. We must therefore ask, under what conditions do all orbits pass through the sticking region?

Fact: If the inverse image and the forward image of $\ominus$ (or $\oplus$) do not intersect in D, then all orbits will pass through the sticking region.

Proof: Suppose $P^-(\ominus) \cap P^{+^{-1}}(\oplus) = \emptyset$. If a point $r \notin P^{+^{-1}}(\ominus)$, and $r \in \oplus$, then $r_1 = P^+(r) \notin \ominus$. Since $P^-(\oplus) \cap \oplus = \emptyset$, $r_1 \in R$. On the other hand, if $r \in P^{+^{-1}}(\ominus)$, and $r \in \oplus$, then $r_1 = P^+(r) \in \ominus$, and $r_2 = P^-(r_1)$. By the hypotheses that $P^-(\ominus) \cap P^{+^{-1}}(\oplus) = \emptyset$, we know that $r_2 \notin P^{+^{-1}}(\ominus)$. Since $r_2 \notin P^{+^{-1}}(\ominus)$, we have seen in the above argument that $P^+(r_2) \in R$.

In other words, if we follow any point s starting in $\ominus$, its mapping $r = P^-(s)$ will be in the image of $\ominus, P^-(\ominus)$. If $P^-(\ominus)$ does not intersect the preimage of $\ominus$, then r is not in the preimage of $\ominus$. If r is not already in R, then its mapping $P^+(r)$ will certainly be in R, and the motion will stick. In the case when the initial $r \in R$,

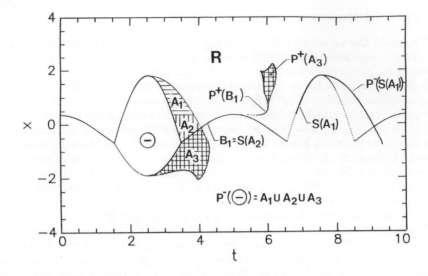

Figure 16: Successive mappings of $\ominus$. Within one period of excitation, the entire set of points has condensed to a line. $\Omega = 1.25, a = 1.9$, and $k = 1.5$.

the motion begins trivially in the sticking region.

This fact provides a sufficient condition for all trajectories to eventually pass through the sticking regions. We need no proof for the images and preimages of $\oplus$ because all orbits in $\oplus$ either map to R or to $\ominus$.

It can be shown numerically that, for the parameters of focus, $P^-(\ominus)$ and $P^-(\ominus)$ do not intersect[6].

The sequence of mappings of the boundary of $\ominus$ can be viewed as a one-dimensional tent-like map. Depending upon the degree of stretching and folding, we may have periodic or chaotic dynamics. The dynamics of this oscillator has been shown to be associated with a one-dimensional single-humped map (see section 4.2). Unlike the usual case where the one-dimensional map is an approximation which exploits a strong stable foliation, this one-dimensional map arises exactly.

5.3. Construction of the Attractor

The sequence of mappings of B^- can be viewed by wrapping $t(\mathrm{mod}\ \frac{2\pi}{\Omega})$ around and back to itself (in $\mathbf{S}^1$). These mappings can then be extrapolated into a flow in $(x, \dot{x}, t(\mathrm{mod}\ \frac{2\pi}{\Omega}))$-space. The resulting template (Figure 17) resembles the images of Figures 3, 4, and 8, displayed in the same space. The dynamics is in a branched manifold. Whether the branched manifold produces a strange attractor or a periodic

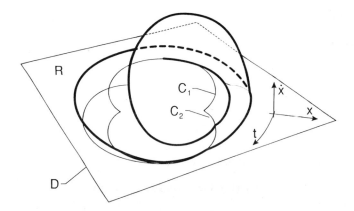

Figure 17: The attractor is constructed by identifying periodicity in the time variable, and extrapolating the previous map sequence into a flow.

attractor depends upon the parameter values. Whether all the motions go through the sticking region also depends upon parameter values.

Templates of the Lorenz attractor[10] and the Rössler attractor[24] are prime examples of branched manifolds. (Guckenheimer and Holmes[10] demonstrate the construction of the Lorenz branched manifold, and further geometric analysis.) Their motions are strongly asymptotically attracted to the branched manifolds. The semiflow on the branched manifold is an approximation, and does not fully represent the detailed behavior of the system.

In the Coulomb oscillator, however, the motion is condensed suddenly (not asymptotically) onto the branched manifold. This noninvertible condensation, due to the discontinuity and multivaluedness of the Coulomb friction law, gives the Coulomb oscillator the opportunity to have a strange attractor in a two-dimensional manifold. The dynamics in the branched manifold fully represents the long-term behavior of the system.

When analyzed using constrained equations, chaos in Lorenz- or Rössler-like attractors is also confined to a two-dimensional manifold[47]. The forced belt-driven oscillator[17,8] should also be confined to a two-dimensional manifold.

The entire attracting set for this system consists of the attracting branched-manifold, and of the permanent-sticking region which surrounds this toroidal structure. The permanent-sticking region exists on D for all x_1 greater than the maximum of $\frac{-1+a}{1+k}$ and $\frac{1-a}{1-k}$. These values represent the maximum values of curves C_1 and C_2. All points landing in this part of the sticking region are stuck forever.

The dynamics on branched manifolds may be viewed via one-dimensional maps, i.e. maps of the form $s_{n+1} = y(s_n)$. A map that would account for both the dynamics on the branched manifold, and the permanently sticking motions, would consist of a component ($s_n \geq 0$) resembling a single-humped map, and a component ($s_n < 0$) coinciding with the identity line, respectively. The component coinciding with the identity line produces an infinite locus of fixed points for all $s_n < 0$. Motions on $s_n > 0$ are dynamic, and may have periodic or chaotic attractors. Motions that get mapped to $s_n < 0$ are trapped there forever.

5.4. The Limit of Large Excitations

Let us look at the geometry as the excitation amplitude a becomes large. Letting $x = az$, the approximate equations for large a are given as

$$\ddot{z} + (1 + k)z = \cos \Omega t, \qquad \dot{z} \geq 0, \tag{15}$$

$$\ddot{z} + (1 - k)z = \cos \Omega t, \qquad \dot{z} \leq 0, \tag{16}$$

$$\ddot{z} + z = \cos \Omega t, \qquad z < 0.$$

Hence, as $a \to \infty$, the constant term of each piecewise continuous system becomes insignificant; the system approaches piecewise linearity. Therefore, the form of the mapping of each boundary in Figure 16 approaches a limiting form.

With the scaled coordinate z, the curves defining the sticking regions approach the form

$$C_1 : z = \frac{\cos \Omega t}{1 + k},$$

$$C_2 : z = \frac{\cos \Omega t}{1 - k},$$

as $a \to \infty$. This is true for all $k \neq 1$. The ratio of the amplitudes of these curves is $r = (1 + k)/(1 - k)$. The parameter value $k = 1.5$ leads to $r = 5$. For the case of $k = 1$, the curves defining the sticking regions are vertical lines. Hence, the sticking regions are vertical strips.

Thus, the geometry of the sticking regions appoaches a limiting geometry in the same scaled coordinate z for which the mappings of the boundaries approach a limiting form. Thus, the entire system dynamics approach a limiting dynamics as $a \to \infty$. This will qualitatively be of the form of the branched manifold constructed in the previous section. This limiting dynamics is in the scaled coordinate z. In the real coordinate x, the dynamics grows with a. System parameters determine whether all trajectories exhibit sticking behavior, whether the attractor is chaotic, or whether all motions go into the permanent-sticking region.

In the case of $k = 0$, C_1 and C_2 coincide, and equations (15) and (16) are the same. Thus, the system approaches the behavior of a frictionless system.

5.5. *Existence and Uniqueness of Solutions*

Existence and uniqueness for systems with simple discontinuous friction laws has been addressed in various contexts. References to this topic go back to Painlevé[48] in 1895. With robot linkages and sliding bars with coulomb friction ($f = \mu sign(v)$), the normal load, and hence the friction, is dependent on the acceleration. When the Coulomb model is used, such systems have been shown to have the possibilities of both multiple solutions and no solutions[49,50,51].

The model studied here does not have the influence of acceleration on the frictional force. Existence and uniqueness of such vector fields has been studied by Filippov[52]. Our concern is actually with the choice of the friction model, namely in modeling the discontinuity. The geometric discussions above lead to a simple criterion on whether the discontinuous function will admit solutions.

The multivaluedness of $f(x_2)$ at $x_2 = 0$ is necessary for the existence of solutions in the sticking region. If f were a single-valued function, and, say, $f(0) = 0$, what would happen? A trajectory in the sticking region in D would be governed by the equations (1) and (2) with $f(0) = 0$. This reduces to a case where there is no static friction, and has the form of equation (12). An orbit in D would be, with probability one (unless it lies on the curve defined by $x_1 = a \cos \Omega t$, in which case it would be in equilibrium for an instant, and then no longer be on the curve), directed out of D. Such a trajectory, if it were to exist, would have to immediately leave D. But it cannot leave D since the flow outside D is directed toward D. So the trajectory does not exist. The multivaluedness of f provides a bridge for the flow across D through the sticking region. The discontinuous $f(v)$ can have a multivalued bridge which exceeds the connection between $+1$ and -1 ($\mu_s > \mu_k$). Then the sticking regions are larger, but trajectories can still remain directed in D while in the sticking region.

Thus, solutions to equation (1) subject to the commonly and loosely defined friction function $f(\dot{x}) = sign(\dot{x})$ will not exist in the neighborhood of regions in the discontinuity where the sense of the adjacent vector fields are in conflict. However, in defining a multivalued friction law through the ideas of static friction, solutions exist everywhere.

Generally, the flow as defined on the discontinuity must be compatible with the piecewise flows adjacent to the discontinuity. Filippov's solution construction ensures such compatability[52].

The numerical procedure outlined in section 4.2.2 effectively implements the existence criterion by checking for sticking, and evaluating the time at which slip begins. Such an algorithm assumes that the path of the trajectory is well defined from stick until slip.

Trajectories described in this geometrical setting are unique in the forward sense. We have already pointed out that sticking trajectories are noninvertible. This could be interpreted as nonunique in backward time.

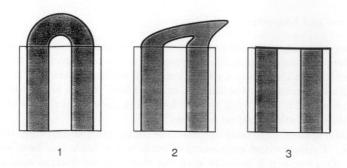

Figure 18: A modified horseshoe map includes condensation of the bent part into a line. This line is then reinjected into the map. The attractor has a portion of fractal dimension $1 + \frac{\ln 2}{\ln 3}$, interlaced with a portion of dimension one.

5.6. Nonuniform Dimension and Topology

We know that stick-slip can lead to collapse in the dimension of phase space[4,35,6]. However, once the dimension has collapsed, can the attracting dynamics reestablish the higher dimensional character?

Suppose we were to have nonlinearities in addition to the friction, and the mappings P^+ and P^- were to correspond to nonlinear flows. Imagine that the mapping of $\oplus$ were to stretch and fold onto $\ominus$ like a horseshoe, with some parts overhanging into R. A schematic is a modified Smale horseshoe map[53] shown in Figure 18. The part of the image normally lost from the square and thereafter ignored is instead compressed to a line, as in the sticking region, and reinjected into the map. We still have the fractal invariant set Λ of dimension d_Λ of the Smale horseshoe, in addition to a set of dimension d_s corresponding to iterations of the map of the compressed reinjected lines. All orbits will end up on one of these two sets. Infinite iterations of the reinjected line will also produce a set of dimension d_Λ and a set of dimension d_s. Thus, if an orbit passes through the squashed line, it passes through a one-dimensional object. With further iterations, it may approach an object of dimension d_Λ. The composite attractor contains components of dimension one interlaced with a component of dimension d_Λ. The two components have different dimension and topology.

Conceivably, such events might take place in a flow if the dynamical system were to have a sticking region, and a horseshoe which leaves portions of its image in the sticking region. In attempting to cook up such an example, we modify the Duffing equation, which has horseshoes, by adding a localized dry-friction damper. The hope is that this example produces a flow a fractal set due to the stretching and folding of the horseshoe, in conjunction with a non-fractal set due to the smashing that takes place in the sticking region.

Figure 19 shows a Poincaré map of a two-well oscillator with an applied dry damper. The equation for this oscillator is

$$\ddot{x} + .1\dot{x} - x + x^3 + F(x, \dot{x}) = A\cos 2.1t, \qquad (17)$$

where $F(x, \dot{x})$ represents the friction force, and $A = 1.2$. The friction function emulates a damper located roughly in the range $x \in [-0.25, 0.25]$ by the function $F(x, \dot{x}) = n(x)f(\dot{x})$, where

$$n(x) = \frac{1}{2}(\tanh(50(x - \frac{1}{4})) - \tanh(50(x + \frac{1}{4}))),$$

and $f(\dot{x})$ is defined as in equation (2). Physically, in this system, motions can momentarily stick in the damper and experience condensation, or they can oscillate freely near either well, or pass between wells without sticking, and undergo folding as in the Duffing system. The Poincaré section appears to have regions where the dimension is one, off which branch segments of higher dimension. Since iterations of a one-dimensional region are mixed with the higher-dimensional portion, it is difficult to separate regions and calculate correlation dimensions. For a Duffing oscillator with a damper approximated by a continuous friction law, and $A = 1.1$, we were able to obtain the correlation dimension on portions of its Poincaré map. A calculation on a selected set of points from its Poincaré section yields the correlation dimension on a region of the section. We calculated a correlation dimension of 0.98 on the central region which looks like a heavy, crisp line, and a correlation dimension of 1.3 on the right and left lower lobes of the section.

Because of the Coulomb damper, true sticking motion takes place, accompanied with condensation of orbits. The low-dimensional portion of its Poincaré section in Figure 19 is presumably not Cantored, with a capacity of one. In such case, an orbit passes through portions of the attractor with varying dimension and varying topology (Cantored and not Cantored). Neighborhoods of an orbit experience an evolution in the topology of the surrounding attractor.

This is preliminary, and a more careful quantification of nonuniform topologies in the attractor is currently under consideration.

5.7. Convergence of Limit Sets

This section has emphasized the matching of piecewise flows from a geometric viewpoint. Another possible route for studying nonsmooth systems might be in examining smooth systems which become nonsmooth in some limit[41].

If we integrate equation (17) with a smooth friction function, e.g. $\hat{f}(\dot{x}) = \tanh(\alpha\dot{x})$, with $\alpha = 50$, the Poincaré section (Figure 20) looks strikingly similar to that of the discontinuous vector field. We suspect the same trend would take place based for Coulomb and smooth models of the mass-spring system based on results from section 4.

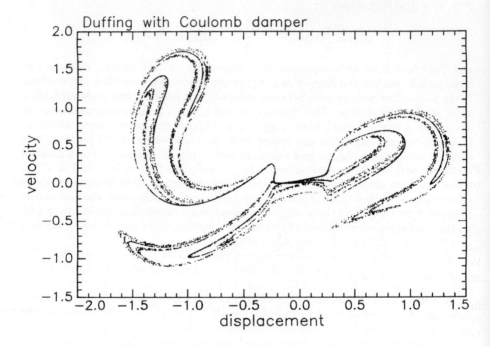

Figure 19: A Poincaré section taken as a slice in time of a Duffing oscillator with friction applied when $-.25 < x < .25$. The coefficient of friction is represented by the Coulomb law with $\mu_s = \mu_k$.

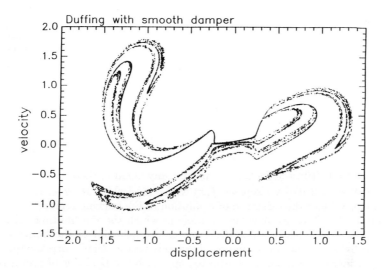

Figure 20: A Poincaré section taken as a slice in time of a Duffing oscillator with friction applied when $-.25 < x < .25$. The friction is modeled by $\tanh(50\dot{x})$.

The smooth damper does not produce true sticking motion, since the lack of a multivalued discontinuity does not produce sticking regions. There is no condensation (smashing is not complete) when the motion is "almost sticking." Thus, the low-dimensional portion of the Poincaré section of Figure 20 is actually a Cantored set of dimension slightly greater than one.

We would expect that as the parameter α increases, and the smooth friction function $\hat{f}$ better approximates the discontinuity, that the attractors in Figures 19 and 20 would become more similar. This leads us to consider sequences of vector fields,

$$\dot{\mathbf{x}} = \mathbf{g}^n(\mathbf{x}, t),$$

where $\mathbf{g}^n(\mathbf{x}, t) \to \mathbf{g}(\mathbf{x}, t)$ in some way as $n \to \infty$. It seems that for certain convergent sequences of vector fields, the sequence of ω-limit sets converges. (It is important to look for the convergence of ω-limit sets, and not the convergence of solutions. If the limiting ω-limit set has a strange attractor, then two orbits which are not identical will generally separate because of a positive Lyapunov exponent.)

If the ω-limit sets of the sequence of vector fields converge, the result may not represent the ω-limit set of the vector field

$$\dot{\mathbf{x}} = \mathbf{g}(\mathbf{x}, t),$$

using a usual idea of convergence. In fact, if $\mathbf{g}(\mathbf{x}, t)$ is not Lipschitz, it could be that no solution exists over the region of **x** to be considered.

In the Duffing oscillator with an applied dry damper, we can approximate the Coulomb law $f(\dot{x})$ with $\hat{f}(\dot{x}) = \tanh(\alpha\dot{x})$, and obtain good numerical results. Similarly, using $\hat{f}(\dot{x})$ in equation (1) also yields good numerical results, as did the smooth law used in section 4.3. As $\alpha \to \infty$, the function $\hat{f}(\dot{x})$ converges to

$$\bar{f}(\dot{x}) = 1, \qquad \dot{x} > 0,$$

$$\bar{f}(\dot{x}) = 0, \qquad \dot{x} = 0,$$

$$\bar{f}(\dot{x}) = -1, \qquad \dot{x} < 0,$$

in the sense that $\hat{f}(0) = 0$ for any y, and for any small ϵ, there is an α such that $|\hat{f}(y) - \bar{f}(y)| < \epsilon$. But if we replace $f(\dot{x})$ with $\bar{f}(\dot{x})$ (which is equal to $\mathrm{sign}(\dot{x})$) in equation (1), we lose the existence of a solution in the sticking region, as discussed in section 5.1.1. Thus, we have an example in which the the limiting behavior of a sequence of vector fields is not the same as the behavior of the limiting vector field. The behaviors of the sequence of smooth vector fields, corresponding to a sequence of increasing α values, (seem to) approach the behavior of a multivalued vector field, which, in the usual sense, is not the limiting vector field.

5.8. Summary

This section presented a geometric way of describing the dynamics of a discontinuous, multivalued Coulomb oscillator. The technique is a three-dimensional extension of that previously employed in classical texts on two-dimensional autonomous systems.

The analysis reconstructs an attractor similar to that seen in numerical integrations. During the reconstruction, it was shown that infinitely strong contraction takes place during sticking motion. Because of this condensation, a one-dimensional map can describe the long-term dynamics exactly. This is unlike the usual case in which a one-dimensional map is an approximation which exploits a strong stable foliation.

Some other properties of the behavior which result from the discontinuous and multivalued nature the vector field (the same mechanism which gives rise to stick-slip motion) are

- the flow may not be invertible

- the flow may reach the attractor in finite time

- strange attractors may have dimension less than or equal to two

While there may be some debate over whether discontinuities really exist in physical systems, certainly *near* discontinuities exist, and to the resolution of measurements, they may be indistinguishable from actual discontinuities.

It seems that oscillators undergoing stick-slip have the potential for having strange attractors with nonuniform topology, in the sense that a trajectory's neighborhood in an attractor might vary in its topology while the orbit flows. The case examined briefly here suggested that the orbit could traverse through an attractor which is locally Cantored in some regions, but not in others.

While the geometric perspective on matching piecewise-smooth flows was emphasized, it is also conceivable that one might examine nonsmooth systems as a limit of a sequence of smooth systems. Visual evidence suggests that the sequence of limit sets of smooth systems can converge to the limit set of a nonsmooth system. However, the vector field itself might converge to a subtly different nonsmooth vector field. Issues such as existence might be heeded.

6. Symbol Dynamics

Chaotic motion in maps can be characterized by symbol sequences. Since stick-slip motion can reduce a 3-D flow to a 1-D map, it may be worthwhile to consider symbol dynamics. Stick-slip motion provides a natural basis for producing symbol sequences during the motion, where, for example, S represents motion which is sticking, and N represents motion which is not sticking (slipping). The resulting symbol sequence can be used to characterize the dynamics. In this section, we apply symbol dynamics to characterize chaos using binary autocorrelation functions and macroscopic Lyapunov exponents[54]. We continue the study by comparing the bifurcation sequence, and its associated symbol sequences, of the Coulomb oscillator, with the "universal" bifurcation sequence of "standard" one-dimensional maps[55]. Universal behavior refers to behavior that is consistant for all parameter ranges in a given class of systems[9].

6.1. *The Binary Autocorrelation and Macroscopic Lyapunov Exponent*

Singh and Joseph[54] have proposed a technique for extracting quantitative information from a binary symbol sequence. First it is necessary to represent the symbol sequence $u(k)$ as a string of 1's and -1's. These values are chosen so that the expected mean of a random sequence of equally likely symbols is zero. As the trajectory passes through the Poincaré section for the kth time, if it is not sticking, we set $u(k) = 1$. If it is sticking, we set $u(k) = -1$. A binary autocorrelation function on such a symbol sequence is defined as

$$r(n) = \frac{1}{N} \sum_{k=1}^{N} u(k+n)u(k),$$

for $n = 0, 1, 2, \ldots$, and $N >> n$. If the sequence is chaotic, the autorcorrelation should have the property $r(n) \to 0$ as $n \to \infty$.

If the sequence becomes uncorrelated, an estimate of the largest Lyapunov exponent can be obtained by using the binary autocorrelation function. The largest

Lyapunov exponent can be defined as

$$\lambda = \frac{1}{N} \sum_{n=1}^{N} \log_2 \frac{d(n)}{d_o(n-1)}, \tag{18}$$

where $d_o(n-1)$ is the difference between two nearby trajectories at the $(n-1)$th iterate, and $d(n)$ is the distance between them after one iteration. Since the binary sequence is uncorrelated, we can estimate $d_o(n-1)$ as the expected distance $\bar{d}_o(n-1)$ between two randomly chosen points in the same symbol region. In our example, we measure the distance using coordinate s on the Poincaré plot. Two points chosen from the sticking region have an expected distance $\bar{d}_o(n-1) = 1/3$. Two points from the nonsticking region have the same expected distance. If $u(n-1)$ and $u(n)$ are in the same region, their iterates will either stay in that region, be in different regions, or both be in the other region. One defines

$$\alpha = \log_2 \frac{\bar{d}(n)}{\bar{d}_o(n-1)},$$

where $\bar{d}(n)$ is the expected distance between two points when each is chosen from separate regions. For our problem, again using s to measure distances, $\bar{d}(n) = 1$ and $\alpha = \log_2 3$. Replacing $d_o(n-1)$ and $d(n)$ in equation (18) with their expected values defines the *macroscopic* Lyapunov exponent, λ_m. The exponent is dubbed *macroscopic* because the distances involved are not infinitesimal. Through the derivation of Singh and Joseph[54], the macroscopic Lyapunov exponent can be written in terms of the binary autocorrelation function:

$$\lambda_m = \frac{1}{2}\alpha[1 - r(1)^2].$$

Application of these ideas to a symbol sequence derived from the tent map yields a rapidly decaying autocorrelation and a Lyapunov exponent $\lambda_t = 0.787516$ for a string of 100000 symbols, and an exponent of $\lambda_t = 0.787705$ for a string of 2048 symbols. Its exact "microscopic" exponent, calculated using $\log_2$, is $\lambda_{te} = 1$. While the tent map has uniform stretching in the sense that the slope has a magnitude of two for all $x \in [0, \frac{1}{2}) \cup (\frac{1}{2}, 1]$, the difference between the microscopic and macroscopic values comes from the fact that the global behavior of the map includes folding. Application to the logistic map yields a rapidly decaying binary autocorrelation function, and a macroscopic Lyapunov exponent of $\lambda_l = 0.791578$ for a sequence of 100000 symbols, and $\lambda_l = 0.791116$ for 2048 symbols, compared to an exact microscopic exponent of $\lambda_{le} = 1$.

Binary sequences of length 2048 were obtained from the experimental oscillator of section 3 and from a numerical simulation using the smooth friction law of section 4.3. The macroscopic Lyapunov exponent for the experimental oscillator was calculated to be $\lambda_{exp} = 0.79055$. That of the smooth-law simulation was computed as $\lambda_{s1} = 0.79219$. In section 4.3, the largest microscopic Lyapunov exponent for the smooth-law flow was estimated numerically[38]. It can be related to that of the Poincaré map via $\lambda_{flow} = \lambda/T$, where T is the driving period. This calculation of the exponent for the Poincaré map from the equations of motion produced $\lambda_{s2} = 0.77$.

The actual computation of the macroscopic Lyapunov exponent can be carried out for symbol sequences from either a deterministic or random source. Thus, it could not be used to distinguish chaos from noise. However, using symbol dynamics thusly can lead to an estimate of the order of magnitude of the largest Lyapunov exponent. In the case of the friction oscillator, such a calculation can be based on data concerning stick and slip. Such 'yes' and 'no' data could conceivably be obtained, for example, by microphone.

6.2. The Bifurcation Sequence and Universality

One-dimensional maps of a "standard" form undergo a "universal" bifurcation sequence when the bifurcation parameter is varied. However, the map arising from the Coulomb model does not have "standard" form. Nonetheless, in this section the bifurcation sequence of the Coulomb model is compared to that of the standard one-dimensional maps to see if it exhibits "universal" behavior. All observed components of the bifurcation sequence fit the universal sequence, although some universal events are not witnessed.

From a mathematical viewpoint, universal behavior of standard maps has been studied in detail. Thus, if a dynamical system exhibits universal behavior, then much is already known about the system. In terms of system identification, examining whether behavior is universal might give a clue as to whether the unknown system fits a standard class of systems. On the other hand, for parametric system identification, where differences in behavior with respect to system parameters is of importance, interest might be focused on nonuniform behavior.

6.2.1. Bifurcations of one-dimensional maps

One-dimensional maps of the type

$$x_{n+1} = \lambda g(x_n), \tag{19}$$

where the function $g(x)$ satisfies certain assumptions, have been studied extensively[9,56,57,58,10]. When considering universal sequences of periodic orbits, the critical assumptions are that

1. $g(0) = g(1) = 0$,

2. $g(x)$ is smooth with a unique maximum at x_0, and $\lambda > 0$, and

3. $g(x)$ has a negative Schwartzian derivative for $x \in [0,1] - \{x_0\}$,

where the Schwartzian derivative for a function $g(x)$ is defined as

$$D_s g(x) = \frac{g'''(x)}{g'(x)} - \frac{3}{2}\left(\frac{g''(x)}{g'(x)}\right)^2.$$

For discussions on metric universality, i.e. Feigenbaum numbers, we can relax the above assumptions, and only assume $g''(x_0) < 0$.

The dynamics of maps which satisfy assumptions 1-3 undergo a universal sequence of bifurcations, where the bifurcation parameter is λ. For $\lambda = \lambda_1$ sufficiently small, such that $\bar{x} = \lambda_1 g(\bar{x})$, and $|\lambda_1 g'(\bar{x})| < 1$, $\bar{x}$ is a stable periodic point of $\lambda g(x)$. As λ increases, a periodic cycle remains until $\lambda = \lambda_2$, at which the periodic point loses stability and a stable periodic cycle of period two is born. A stable cycle of period two exists until $\lambda = \lambda_4$, where the period two loses stability and a period four is born. This period-doubling sequence continues, producing stable periodic cycles of period $2^n, n \to \infty$, as λ approaches a limiting value λ_∞. Given any λ such that $\lambda_\infty < \lambda < \lambda_c$, there exists an infinite number of unstable periodic cycles and a stable cycle of period n. The stable cycle undergoes a similar period-doubling sequence as above, to a limiting value of λ_{n_∞}. Typically, on a bifurcation diagram, windows of relatively low period n are identifiable to the eye.

The bifurcation sequence of the map (19) exhibits universal behavior, that is behavior common to any function $g(x)$ which satisfies the assumptions stated above. As the bifurcation parameter increases, there will be a bifurcation sequence of stable periodic orbits. From this sequence, we could construct a sequence of the period lengths of these stable cycles. It is a universal property that this sequence of period lengths is the same for all such maps. For each stable periodic cycle, there exists a parameter value $\hat{\lambda}$ such that one point p_0 of the periodic sequence lies at x_0. The value of a periodic point $p_0(\lambda)$ is continuous in λ, and there is a region $(\hat{\lambda} - \delta_1, \hat{\lambda} + \delta_2)$, for some $\delta_1 > 0, \delta_2 > 0$, such that the periodic cycle remains stable, and p_0 stays near x_0[57].

A symbol sequence for the periodic cycle can be defined by whether the ith iterate in the cycle is to the right of x_0 (R) or to the left of x_0 (L). Since p_0 is arbitrarily close to x_0, we assign the symbol C to p_0. For a cycle of period m, the remaining $m - 1$ iterates of p_0 are assigned the symbols R and L. For example, if the periodic cycle had a period of five, the symbol sequence might be CRLLL, pertaining to iterates of the point located very near x_0. The convention in the literature[57,58] is to drop the symbol C. Therefore, a symbol sequence for a periodic cycle of period m is defined as the symbols of the $m - 1$ iterates of p_0. In our example, the period-five symbol sequence is defined by the four symbols RLLL. A second universal property is that the symbol sequences of each of these stable periodic orbits is the same for all such maps.

Metric universality exists for maps which merely satisfy $g''(x_0) < 0$. In this case, the period-doubling sequence occurs according to Feigenbaum's ratio[9,56]. If λ_n is the parameter value for a cycle of period $m2^n$, then Feigenbaum's ratio is

$$\delta = \lim_{n \to \infty} \frac{\lambda_{n+1} - \lambda_n}{\lambda_{n+2} - \lambda_{n+1}} = 4.6692\cdots.$$

6.2.2. The one-dimensional map underlying the Coulomb-friction oscillator

In section 4.2.3, we saw how a one-dimensional map arised from a return map on a variable s defined on the Poincaré image of the oscillator with Coulomb friction. Let us call this one-dimensional map $F(s)$.

We also saw, in section 5.1.1, how sticking regions can be described by observing the directions of the piecewise continuous vector fields at the discontinuity D, defined in the state space by $\dot{x} = 0$. When both vector fields agree to flow through D, the flow may pass through the discontinuity. When both vector fields point toward D, there is a stable sticking region R. Flows are trapped in R until time evolves such that the orbits are on either of the boundaries, B^+ or B^-, of the sticking region R. A map describing this action would be $S : R \to B^+ \cup B^-$. S is singular since it takes a two-dimensional region R and maps it into a finite union of curves.

For the parameter case studied ($a = 1.9, \Omega = 1.5$, and $k = 1.5$), the entire two-dimensional region $\ominus$ collapses into a one-dimensional curve. This geometrically illustrates the singularity which produces the one-dimensional map $F(s)$. The singularity is in the mapping S in the sticking region. All motions pass through the sticking region.

When all motions pass through R, knowledge of the mappings of the boundaries B^+ and B^- is sufficient to understand the attracting set.

An analytical expression for $F(s)$ would consist of three components. One component would involve orbits passing through the boundary B^- and their intersection with the plane D, represented by $P^-(B^-)$. Finding $y = P^-(x)$ requires the solution of transcendental equations. The mapping $P^-(B^-)$ lies partly in R and partly in $\oplus$. Thus, the second component of the analytical expression of the one-dimensional map would be a logical operation. The third component would then be to either solve for the time at which trajectories in R leave the sticking region at B^+ or B^-, or else to solve the transcendental equation representing those orbits which map to D via P^+.

In short, the analytical description of $F(s)$ is compounded with transcendental equations and logical operations. Because of this complexity, we do not produce such an analytical expression, and our work is done largely from a geometric standpoint.

The map $F(s)$ (Figure 9) may not satisfy all the assumptions in the above discussions. Perhaps most importantly, $F(s)$ does not fit the form of equation (19). Varying a in equation (1) actually alters the resulting shape function g as well as the magnitude λ. This is because the orientations of the sticking-region boundaries (section 5) are dependent on a.

The "instantaneous" shape function g satisfies the first assumption (if the coordinate s is rescaled), but not necessarily the second (taking "smoothness" in the context of the Schwartzian derivative, i.e. up to three derivatives) and third.

Although the map $F(s)$ (and hence g) cannot be determined explicitly, implicit relationships make it possible to calculate derivatives. This is precisely what Shaw did when calculating stabilities of periodic motions in a similar oscillator[4]. To this end, we refer to Figure 16. As an orbit goes through the one-dimensional map, it can undergo one of three series of compositions.

Case 1. Imagine an initial point, (x_0, t), in region A_1, as it evolves until time t_0 at the boundary $S(A_1)$ of the sticking region. Note that x_0 and t_0 are directly related. The corresponding flow goes through negative velocity and returns to the sticking region at time t_1, in the region $P^-(S(A_1))$, which can be determined by solving the transcendental equation $\dot{x}^-(t_1; t_0) = 0$, where $x^-(t_1; t_0)$ represents to the x component of the flow corresponding to P^-, and is the solution to equation (11). If $\frac{\partial x^-}{\partial t_1} \neq 0$, then t_1 is implicitly a function of t_0, i.e. $t_1 = p(t_0)$. Finally, $x_1 = x^-(t_1; t_0)$ completes the one-dimensional mapping.

Case 2. Trajectories in Figure 16 based at some point (x_1, t_1) in region A_2 leave the sticking region at time t_2 in region $B_1 = S(A_2)$. Flowing with positive velocity, the orbit returns to the sticking region at time t_3, in region $P^+(B_1)$, at a value of x_2, which can be determined from the transcendental equation $\dot{x}^+(t_3; t_2) = 0$, where $x^+(t; t_2)$ represents the x component of the flow corresponding to P^+, and is the solution to equation (10). This stuck trajectory leaves the sticking region on the boundary $S(A_1)$, and flows through negative velocity back to the sticking region in $P^-(S(A_1))$, completing the mapping at a point x_3 and time t_4, which can be determined by the transcendental equation $\dot{x}^-(t_4; t_3) = 0$. Then, $x_3 = x^-(t_4; t_3)$.

Case 3. Some motions can be based at a point (x_1, t_1) in region A_3. Here, the point (x_1, t_1) lies in the slipping region, and continues to flow in the same manner as (x_2, t_3) does in the case above. The sticking phase between times t_1 and t_2 does not occur.

Orbits based in region A_3 can involve complications due to the condition of $n(x) = 0$ for values of $x < -1/k$.

These possibilities constitute the 1-D mapping, $F(s)$, for this oscillator. The function $F(s)$ is not expressable in closed form. However we can, in most instances, calculate the derivatives of $F(s)$, and examine the Schwartzian derivative. For example, in Case 1., the map starts at x_0. We can calculate $t_0 = \arccos(x_0(1 - k))/\Omega = \alpha(x_0)$. Then, $t_1 = p(t_0)$, as an implicit function, and the next iterate of the map is $x_1 = x^-(t_1; t_0)$. Dependence on the initial condition x_0 is expressed through t_0. The derivative of the implicit map $F : x_0 \to x_1$ is

$$F' = [\frac{\partial x^-}{\partial t_1}\frac{dp}{dt_0} + \frac{\partial x^-}{\partial t_0}]\frac{d\alpha}{dx_0}, \tag{20}$$

where $\frac{dp}{dt_0} = -\frac{\partial \dot{x}^-}{\partial t_0}/\frac{\partial \dot{x}^-}{\partial t_1}$. F' can be regarded as a function of t_1, t_0, and x_0. In this description, subsequent derivatives have the form

$$F^{(n+1)} = [\frac{\partial F^{(n)}}{\partial t_1}\frac{dp}{dt_0} + \frac{\partial F^{(n)}}{\partial t_0}]\frac{d\alpha}{dx_0} + \frac{\partial F^{(n)}}{\partial x_0}.$$

The first derivative of F for Case 2. is

$$F' = [\frac{\partial x^+}{\partial t_3}\frac{\partial q}{\partial t_2} + \frac{\partial x^+}{\partial t_2}\frac{\partial r}{\partial t_1}]\frac{\partial r}{\partial x_1}[\frac{\partial x^-}{\partial t_1}\frac{dp}{dt_0} + \frac{\partial x^-}{\partial t_0}]\frac{d\alpha}{dx_0}, \tag{21}$$

where $r(x_1) = \arccos(x_1(1 + k))/\Omega$.

If we consider the transition between mapping as in Case 1., and mapping as in Case 2., we will look at the orbits which pass through the intermediate points (x_2, t_2), with $x_2 = 1/(1 + k)$ and $t_2 = 2\pi/\Omega$. This orbit could be classified as either Case 1 or 2, depending on whether, by definition, the action of $r(x_2)$, and $x^+(t_3; t_2)$ takes place. We can numerically compute the derivatives to show that the Schwarzian derivative is not smooth at this transition. Thus, the 1-D map will not satisfy "standard" hypotheses, and we need not expect to observe a universal bifurcation sequence.

We can also see that the maximum of $F(s)$ is smooth, at least for some values of a. This is relevent to discussions on metric universality (Feigenbaum's number). Visual evidence is in Figure 9.

To discuss the maximum of $F(s)$, we must first locate it. To this end, we look at the image $P^+(B_1)$ in Figure 16. Close inspection indicates that this image has a local minimum in x at $x = z$ for some value of t_z ($t_z \approx 5.5$ in the figure). Additionally, for the given parameters, this point z denotes the minimum value of s in some Poincaré sections. In the Poincaré section at $t = 6.5$, for example, defining a coordinate $\hat{s}$ such that $0 \leq \hat{s} \leq 1$ and $\hat{s} = 1$ at z shows that the point $(z, t = 6.5) \in D$ corresponds to the maximum value of $\hat{s}$. In a Poincaré section at $t = t_z$, the point $(z, t_z) \in D$ represents s_0, locating the local maximum in the underlying map.

Since a small neighborhood $V \in B_1$ of the orbit passing through this critical point is governed solely by a function $P^+(V)$ which maps a curve monotonely increasing in x to a curve with a smooth minimum in x (at $x = z$), the point s_0 represents a smooth maximum in the underlying map.

We should point out that if $P^+(B_1)$ were to lack a local maximum in x, then the point s_0 would correspond to sticking orbits passing through the local maximum of the curve B^+ (Figure 15), and thus s_0 would represent a boundary between two functions active in the one-dimensional map (involving P^+ and P^-). In general, such a map would not have a smooth maximum. Whether this occurs for all values of a is not known.

6.2.3. Bifurcation analysis for the Coulomb oscillator

Since the equation of motion has a discontinuity at $\dot{x} = 0$, the plane in $(x, \dot{x}, t)$-space defined by $\dot{x} = 0$ is a natural place to make a Poincaré section. In this Poincaré section we plot x for the bifurcation diagram shown in Figure 21. The bifurcation diagram includes trajectories which bounce off the underside ($\dot{x} < 0$) of the (x, t) plane. (Some trajectories meet the (x, t) plane from below, stick, and then return below the plane. This corresponds, for example, to motion near the outer edge of the attractor in Figure 8.)

The method used to compare the bifurcation sequence in the Coulomb friction model to the standard one-dimensional maps is as follows: We compute and plot a bifurcation diagram, which has periodic windows. We identify periodic orbits

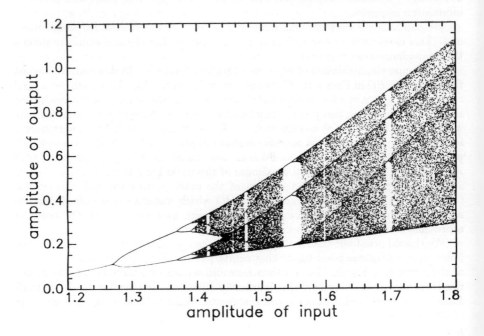

Figure 21: A bifurcation diagram shows period doubling to be the route to chaos. The bifurcation parameter, the driving amplitude a, is increasing in this plot.

visible to a parameter increment Δa of $\Delta a = 0.0005$. In doing so, we look for stable periodic orbits of period less than eight. The infinitely many higher periods are difficult to find because the ranges of a on which they exist are narrow. We compare the sequence of period lengths of stable periodic cycles found in the Coulomb oscillator with the universal sequence of stable periodic cycles. If a period five, say, appears in the bifurcation diagram for a parameter value $a = \hat{a}$, then we observe the Poincaré section (from a slice in time) of the motion with the parameter set to $\hat{a}$. In the Poincaré section the orbit will consist of five points. From this Poincaré section, we can determine the symbol sequence of the periodic points in terms of the sticking and slipping regions. We assign the symbol S to points which are *sticking*, and the symbol N to points which are *not sticking* (slipping). We also compare the symbol sequence of the stable periodic cycles found in the Coulomb oscillator with the universal symbol sequences of stable periodic cycles. Finally, we estimate Feigenbaum numbers from the Coulomb oscillator data.

This sequence of N's and S's will be analogous to the symbol sequence for the associated map, and can be translated into a sequence of R's and L's, respectively. This translation is exact when the Poincaré section is taken at $t(\mathrm{mod}\,\frac{2\pi}{\Omega})$ corresponding to t_z. This is because the maximum of the underlying map occurs at the critical point s_0 corresponding to z, which is the local minimum (in x) of curve $P^+(B_1)$. At a phase corresponding to t_z, this point cleanly separates sticking motions from slipping motions, as well as left from right.

However, when $t \neq t_z$, some other value of $s \neq s_0$ separates points that are sticking from points that are slipping. As an extreme case, when $t = 6.5$ in Figure 16, all points are in the sticking region (a symbol sequence would trivially consist only of S's). In the neighborhood of t_z, the approximation of the N's and S's as R's and L's is reasonable due to the smoothness of the curve $P^+(B_1)$. Deviation of the N's and S's from the R's and L's represents an error in observation of behavior, rather than an effect on universality. Our Poincaré section was taken at $t(\mathrm{mod}\,\frac{2\pi}{\Omega}) \approx 0.625$. Based on Figure 16 viewed at $t \approx 5.6$, error in the symbol dynamics should be small.

6.2.4. Results

A comparison of the sequence of periodic orbits, and their symbol sequences, for the friction oscillator and the logistic map is in Table 1. For the logistic map, $x_{n+1} = \lambda g(x_n)$ with $g(x) = x(1 - x)$. The values for the logistic map were obtained[57,58] for periodic orbits of period seven or less. Two higher-period events were added since they were incidentally found in the friction oscillator.

The bifurcation sequence of a Coulomb friction oscillator, in some sense, resembles the universal sequence of standard one-dimensional maps. The observed periodic orbits, their period lengths and symbol sequences, of the Coulomb friction oscillator lie in the sequence of the standard maps, although several events remain undetected. In other words, every event in the oscillator is also in the universal

Table 1: The observed sequence of period length of stable periodic cycles in the Coulomb oscillator, and their symbol sequences, are compared to the sequence of stable periodic cycles in the logistic map. Each cycle listed is the first of an infinite period-doubling sequence, except those marked with a *, which arise from the previous cycle via period doubling. For the Coulomb oscillator, S indicates points which are sticking, and N indicates points which are not sticking. The symbol • represents a point which was so close to the boundary between N and S that it was not distinguishable. Each periodic cycle has one point which is very close to this boundary. We label such points with the symbol C. The listed symbol sequence of a cycle of period m consists of the $m - 1$ iterates of the point labeled C. Only cycles up to period seven are included in the table. Some of these cycles in the Coulomb oscillator were not found. Incidents of a period eight and a period ten were accidentally found and included.

Osc.	Eqs. (1)-(3)		Log. Map	$x_{n+1} = \lambda x(1 - x)$	
Period	*Symbol Seq.*	*a*	*Period*	*Symbol Seq.*	λ
2	N	1.36	2	R	3.2360680
4*	NSN	1.38	4*	RLR	3.4985617
8*	NSNNNSN	1.3925	8*	RLRRRLR	not listed
10	NSNNNSNSN	1.4064	10	RLRRRLRLR	not listed
6	NSNNN	1.415	6	RLRRR	3.6275575
7	NSNNNN	1.45415	7	RLRRRR	3.7017692
5	NSNN	1.4737	5	RLRR	3.7389149
7	NSNNSN	1.4909	7	RLRRLR	3.7742142
3	NS	1.535	3	RL	3.8318741
6*	NS•NS	1.551	6*	RLLRL	3.8445688
			7	RLLRLR	3.8860459
5	NSSN	1.5973	5	RLLR	3.9057065
			7	RLLRRR	3.9221934
			6	RLLRR	3.9375364
			7	RLLRRL	3.9510322
4	NSS	1.695	4	RLL	3.9602701
			7	RLLLRL	3.9689769
			6	RLLLR	3.9777664
			7	RLLLRR	3.9847476
5	NSSS	1.833	5	RLLL	3.9902670
			7	RLLLLR	3.9945378
			6	RLLLL	3.9975831
			7	RLLLLL	3.9993971

sequence of events (but not *vice versa*), and the order of events in the oscillator does not contradict the order in the universal sequence. Furthermore, the symbol sequences associated with each observed periodic cycle in the Coulomb friction oscillator is the same as in the corresponding periodic cycle in the standard one-dimensional maps.

There are two possible explanations for the fact that some bifurcation events were not detected. One possibility is that these events took place on a parameter window smaller than the resolution at which we chose to search. The other possibility is that the bifurcation sequence of the oscillator in fact may not match the universal sequence of the standard maps. Such deviation could occur since the map which arises from the Coulomb oscillator does not satisfy all of the assumptions required for universal behavior in the standard maps. Thus, it is possible that we are observing a nonuniversal bifurcation sequence in the Coulomb oscillator. Coffman *et al.*[59] previously observed nonuniversal behaviour in nonstandard one-dimensional maps. In that study, the nonuniversal events also occured in the universal sequence, but not according to the universal order.

In the investigation of metric universality, the parameter values in the initial period-doubling sequence were measured to be compared with Feigenbaum's ratio. We measured the parameter value a_2 at the first period-doubling bifurcation, and a_4, a_8, and a_{16} at the subsequent bifurcations. We then compared

$$r_1 = \frac{a_4 - a_2}{a_8 - a_4}$$

and

$$r_2 = \frac{a_8 - a_4}{a_{16} - a_8}$$

to Feigenbaum's ratio of $4.669\cdots$. The estimates are $r_1 = 4.70 \pm 0.25$ and $r_2 = 4.67 \pm 0.65$. In the estimates, the parameter increment was smaller than that of the rest of this investigation. The uncertainty is in the numerical integration. Within the scope of the error, it would not be unusual for the ratio to converge nonuniformly. Higher bifurcations involve smaller window sizes, leading to larger errors in the calculations of the r_i.

6.3. Summary

Symbol dynamics has been used to characterize the dynamics of the Coulomb friction oscillator through binary autocorrelations, macroscopic Lyapunov exponents, and bifurcation sequences.

Variation of a parameter in the continuous-time oscillator does not necessarily correspond to variation of a single parameter in the underlying map. Because of this, and because of other considerations, such as the Schwartzian derivative, the map may not fit the description of the "standard" maps. Nonetheless, the bifurcation sequence of the oscillator has been compared to that of the standard one-dimensional maps.

The observed periodic orbits, their period lengths and symbol sequences, fit into the universal sequence. However, several universal events remain undetected. This may be because the necessary assumptions are not all met, and that the oscillator does indeed exhibit nonuniversal behavior. However, consideration must be given to the size of the increment in the bifurcation parameter, Δa. If a periodic window is smaller than Δa, the window may not be observed. Therefore, a statement regarding the nonexistence of a periodic window cannot be made, since Δa can always be made smaller.

7. Stick-Slip and the Reconstruction of Phase Space

When running experiments, it is not typically feasible to measure all of the active states in an experimental system. To compensate for this, there are methods for estimating the behavior of the full state space from a small number of measured observables. In experimental observation, these methods involve phase-space reconstructions[25]. The reconstruction of the full state space can be combined with other analytical ideas, such as fractal dimensions and Lyapunov exponents, singular systems analysis, and false nearest neighbors, for the purposes of nonlinear prediction, modeling, or simply to estimate bounds on the size of the system[60,61]. However, the methods for reconstructing phase space have been developed for smooth systems.

The phase space reconstruction, is usually done by the *method of delays* (described below). Stick-slip, however, can cause the method of delays to fail! For example, stick-slip can lead to a singularity when mapping the observed time history into a higher dimensional space[6]. This is not surprising since stick-slip is a nonsmooth phenomenon, and the validity of the method of delays has been proven only for smooth phenomena. (Takens' embedding theorem[62,25] states that, if basic hypotheses are satisfied, the method of delays of an observable produces an trajectory in the reconstructed phase space which is diffeomorphic to the trajectory in the real phase space. A stick-slip system does not have the smoothness required for Takens' embedding theorem.)

This failure can be visualized by imagining a discretized stick-slip history with sampled displacements x_n. Suppose, for example, we were to reconstruct the phase space in three dimensions. According to the method of delays, we build vectors (x_n, x_{n+k}, x_{n+2k}). However, during a sticking interval, it is possible that the points $x_n = x_{n+k} = x_{n+2k} = x_{stick}$, and that the two points (x_n, x_{n+k}, x_{n+2k}) and $(x_{n+1}, x_{n+1+k}, x_{n+1+2k})$, for example, might both be the same point as $(x_{stick}, x_{stick}, x_{stick})$. Thus, when we plot the reconstructed vectors, many of them pile up on the identity line. (If k were large, however, x_n, x_{n+k}, and x_{n+2k} could span a time interval greater than the sticking time, and this problem is avoided. However if k is too large, the delayed coordinate may become statistically independent of the reference coordinate.)

In such case, the reconstructed phase space is fundamentally different than the real phase space. The map which takes the real phase-space manifold M_p into the reconstructed manifold M_r is not invertible, and thus not a diffeomorphism. The dimensionality study fails. Furthermore, this kind of study is often performed on systems for which the model is unknown. If a nonsmooth event such as stick-slip occurs without notice, poor results may be unsuspiciously obtained. This translates to inaccurate models and poor characterizations and predictions. For example, correlation integrals did not produce the expected straight-line characteristic in log-log plots versus the box size for the system upon which this chapter focuses. Popp and Stelter have also observed this in the friction-induced chaos of a belt-driven disk[17].

There are two problems to address here. First, we would like to detect when a reconstruction problem of this nature occurs in a "black-box" experiment. Next, when there is a reconstruction problem, we need to make adjustments to relieve it, so that the rest of the dimensionality study may continue.

An example of this problem is illustrated in a numerical study of equation (1) with equations (2) and (3). The result of using the observable x is used to reconstruct the phase space by the method of delays is shown in Figure 22. The region of phase space trajectories between curves AB and CD are sticking. All of these points are collapsed onto the line segment EF (which is in fact on the identity), in the reconstructed phase space, during the action of the reconstruction. Thus, much information is lost.

In this system, we can remedy the situation with a reconstruction based on *two* variables. If we choose x and $t(\mathrm{mod}\, 2\pi/\Omega)$ as the observables, and perform delays on x, the time variable serves to unfold these trajectory points which are otherwise collapsing onto the identity line. Thus no collapse occurs. A computation of false nearest neighbors[60] has a healthy characteristic for the latter case, but not the former (Figure 23).

We can use a near-neighbors method to identify the collapsing problem. In the above examples, we saw how a stick-slip observable can collapse if the delay index k is not too large. Thus, as we increase the delay k, the number of incidences of collapse should decrease. A plot of the number of points whose nearest neighbors lie within some prescribed distance vs. k is shown in Figure 24. The expected result shows that the collapsing events occur for certain intervals of k centered around harmonics of the driving period of the system. This is indicated by the + symbols, where there are large numbers of near neighbors for intervals of k. Adding $t(\mathrm{mod}\, 2\pi/\omega)$ as an additional observable removes this feature, as seen by the o symbols.

We are seeking other tools for identifying this type of event, and will apply them to systems in which the stick-slip is not so visually obvious. The goal is to reveal signs of the event when it might not be recognized visually. For example, singularities in reconstructions are likely to effect redundancy computations as well, since they quantify the "sharpness of probability distributions" in the delay-coordinate space[63]. Iterates of singular points in reconstruction spaces are nonunique. Thus, we expect redundancy analysis to diagnose singularity events in reconstructions.

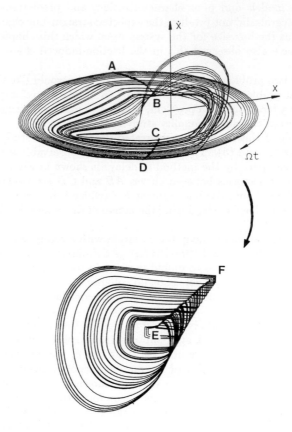

Figure 22: An illustration of the collapse which effectively takes place during a phase space reconstruction. The sticking region between AB and CD collapses singularly into EF.

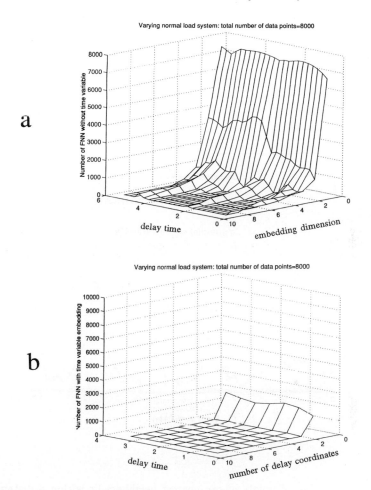

Figure 23: The number of false nearest neighbors vs. the time delay and the embedding dimension for reconstructions with (a) displacement as the only observable, and (b) displacement and time $t(\mathrm{mod}\, 2\pi/\Omega)$ as the two observables. The plot in (a) does not converge since, as the embedding dimension increases, points are removed from the line of collapse, causing them to appear as false nearest neighbors. The plot in (b) has a healthy characteristic since the time observable unfolds the line of collapse in the reconstruction. It indicates that two or three delay coordinates are necessary to unfold the attractor.

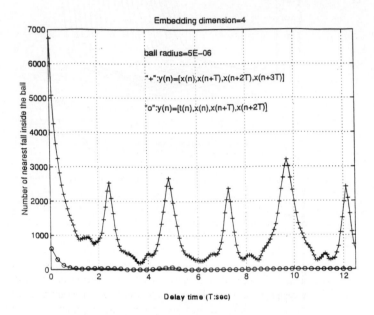

Figure 24: The number of points whose nearest neighbors lie within a distance of $\epsilon = 5e - 6$ for different values of delay τ. When τ is such that there is collapsing in the phase space reconstruction, we have many more near neighbors than otherwise. The + symbols represent single-observable reconstructions. The o symbols indicate the number of points with ϵ-near neighbors for reconstructions from two observables, x and $t(\mathrm{mod}\, 2\pi/\omega)$.

We have also observed that correlation sums do not scale uniformly with the size of the reference balls. Whether they produce a distinctive feature indicative of singularities in the reconstruction is also under investigation.

8. Conclusion

In this chapter, we have studied forced single-degree-of-freedom oscillations with stick-slip friction, mostly through a particular example. An important phenomenon of such oscillators it that stick-slip causes a collapse in phase-space. For a three-dimensional system, this leads to one-dimensional map dynamics. This aspect of the dynamics has been observed experimentally, and numerically with a Coulomb friction model. The phenomenon occurs approximately for smooth friction laws.

The geometry of the flow was studied to show how the dimension collapses when phase flow interacts with sticking regions. With this geometric perspective, it can be seen how the flow is noninvertible. Attractors can be reached in finite time. Chaos can occur in a 2-D manifold. The dimensional collapse seems not to condemn the attractor to a lower-dimensional entity. Nonlinear piecewise-smooth flows can apparently stretch and fold a collapsed volume to an extent which reestablishes a higher-dimensional quality, resulting in a nonuniform topology for the attractor.

Since oscillations with stick-slip can be represented by discrete maps. One-dimensional maps have been studied extensively. Symbol dynamics provides a useful tool. Stick and Slip are natural ways to define a symbol sequence. We have used symbol sequences for this oscillator to compute binary autocorrelations and estimate the order of magnitude of the maximum Lyapunov exponents. We have also examined the bifurcation sequence in terms of symbol dynamics.

Finally, we discussed problems that stick-slip can pose for reconstructing phase space from a sampled observable. A remedy has been presented for the oscillator of focus. The long-range goal is to be able to identify a reconstruction problem in a "black box" nonsmooth system, and also to extend the current reconstruction techniques to accomodate the nonsmoothness.

Stick-slip oscillators represent a class of mechanical systems. The ideas presented in this chapter should shed light on understanding the nonlinear dynamics of such systems. Hopefully, such understanding will facilitate research and development in areas including robotics, automotive squeak, rail-wheel dynamics, micromachines, and earthquake engineering.

9. Acknowledgements

I would like to express great appreciation to F. C. Moon. I would also like to thank others who have had a positive influence on this work, including P. J. Holmes, J. Guckenheimer, T. Kiemel, Bill Feeny, A. Ruina, J. W. Liang, J. P. Cusumano, W. Holmes, and anonomous reviewers.

10. References

1. R. A. Ibrahim, *Friction-Induced Vibration, Chatter, Squeal, and Chaos*, ASME Winter Annual Meeting, DE-Vol. **49** (1992) 107-138.
2. B. Armstrong-Hélouvry, P. Dupont, and C. Canudas de Wit, *Automatica* **30** (7) (1994) 1083-1138.
3. A. Polycarpou and A. Soom, *Friction-Induced Vibration, Chatter, Squeal, and Chaos*, ASME Winter Annual Meeting, DE-Vol. **49** (1992) 139-148.
4. S. W. Shaw, *Journal of Sound and Vibration* **108** (1986) 305-325.
5. F. Pfeiffer, *Zeitschrift für angewandte Mathematik und Mechanik* **71** (1991) T6-T22.
6. B. Feeny, *Physica D* **59** (1992) 25-38.
7. B. F. Feeny and F. C. Moon, *Physics Letters A* **141** (8,9) (1989) 397-400.
8. K. Popp, *Friction-Induced Vibration, Chatter, Squeal and Chaos*, ASME Winter Annual Meeting, DE-Vol. **49** (1992) 1-12.
9. R. L. Devaney, *An Introduction to Chaotic Dynamical Systems* (Addison-Wesley, Redwood City, California, 1987).
10. J. Guckenheimer and P. Holmes, *Nonlinear Oscillations, Dynamical Systems, and Bifurcations of Vector Fields* (Springer-Verlag, New York, 1983).
11. Z. J. Kowalik, M. Franaszek and P. Pierański, *Physical Review A* **37** (1988) 4016-4022.
12. W. Szczygielski and G. Schweitzer, *Dynamics of Multibody Systems*, edited by G. Bianchi and W. Schiehlen (Springer, Berlin Heidelberg, 1986) 287-298.
13. V. I. Utkin, *IEEE Transactions on Automatic Control* **22** (1977) 212-222.
14. W. Eckolt, *Zeitschrift für technische Physik* **7** (1920) 226-232.
15. J. P. Den Hartog, *Transactions of the ASME* **53** (1931) 107-115.
16. B. Feeny and F. C. Moon, *Journal of Sound and Vibration* **170** (1994) 303-323.
17. K. Popp and P. Stelter, *Philosophical Transactions of the Royal Society of London, A,* **332** (1990) 89-105.
18. F. C. Moon, *Physics Letters A* **132** (5) (1988) 249-252.
19. R. Pratap, S. Mukherjee, and F. C. Moon, *Journal of Sound and Vibration* **172** (1994) 321-358.
20. J. M. Carlson and J. S. Langer, *Physical Review A* **40** (11) (1989) 6470-6484.
21. D. P. Vallette and J. P. Gollub, "Spatiotemporal dynamics due to stick-slip friction in an elastic membrane system," preprint (1993).
22. A. V. Srinivasan, *Large Space Structures: Dynamics and Control*, edited by S. N. Atluri and K. A. Amos (Springer-Verlag, Berlin, 1987).
23. B. Feeny, "Chaos and Friction," PhD Thesis, Theoretical and Applied Mechanics, Cornell University (1990).
24. A. V. Holden and M. A. Muhamad, *Chaos,* edited by Arun V. Holden (Princeton University Press 1986) 15-36.
25. N. Gershenfeld, *Directions in Chaos, II*, edited by Hao B.-L. (World Scientific

26. J. T. Oden and J. A. C. Martins, *Computer Methods in Applied Mechanics and Engineering* **52** (1985) 527-634.
27. B. Armstrong-Hélouvry, *IEEE Trans Aut Contr* **38** (1993) 1483-1496.
28. J. R. Anderson and A. A. Ferri, *Journal of Sound and Vibration* **140** (2) (1990) 287-304.
29. C. Pierre, A. Ferri, and E. Dowell, *J Appl Mech* **52** (1985) 958-964.
30. J. J. Stoker, *Nonlinear Vibration* (Interscience publishers, New York, 1950).
31. A. H. Nayfeh and D. T. Mook, *Nonlinear Oscillations* (Wiley, New York, 1976).
32. J. Dieterich, *J Geophysical Res* **84** (1979) 2161-2168.
33. A. Ruina, Slip instabilities and state variable friction laws, *J Geophys Res* **88** (B12) (1983) 10359-10370.
34. A. L. Ruina, *Mechanics of Geomaterials*, edited by Z. Bazant (Wiley, Chinchester, 1985) 169-187.
35. Ch. Glocker and F. Pfeiffer, *Nonlinear Dynamics* **3** (1992) 245-259.
36. J. P. Meijaard, *Fifth Conference on Nonlinear Vibrations, Stability, and Dynamics of Structures*, Blacksburg, June 12-16, 1994.
37. S. F. Bockman, proceedings of the *American Control Conference* **2** (1991) 1673-1678.
38. F. C. Moon, *Chaotic and Fractal Dynamics: An Introduction for Applied Scientists and Engineers* (Wiley, New York, 1992).
39. M. Linker and J. Dieterich, *J Geophysical Res* **97** (B4) (1992) 4923-4940.
40. P. E. Dupont and D. Bapna, *J Vibr Acoust* **116** (1994) 237-242.
41. W. Kleczka and E. Kreuzer, *Nonlinearity and Chaos in Engineering Dynamics,* edited by J. M. T. Thompson and S. R. Bishop (Wiley, Chinchester, 1994) 115-124.
42. A. A. Andronov, A. A. Vitt, and S. E. Khaikin, *Theory of Oscillators* (Dover Publications, New York, 1966).
43. F. Takens, *Lecture Notes in Mathematics* **535** (Springer-Verlag, 1976) 237-253.
44. H. Oka, *Japan Journal of Applied Mathematics* **4** (1987) 393-431.
45. M. Zak, *Neural Networks* **2** (1989) 259-274.
46. S. S. Antman, *Quarterly of Applied Mathematics* **XLVI**(3) (1988) 569-581.
47. H. Oka and H. Kokubu, *Patterns and Waves—Qualitative Analysis of Nonlinear Differential Equations* (1986) 607-630.
48. P. Painlevé, *Comptes Rendus de l'Académie des Sciences* **121** (1895) 112-115.
49. P. Lötstedt, *ZAMM* **61** (1981) 605-615.
50. M. T. Mason and Y. Wang, *Proceedings of the 1988 IEEE International Conference on Robotics and Automation,* 524-528.
51. P. E. Dupont, *Proceedings of the 1992 IEEE International Conference on Robotics and Automation,* 1442-1447.
52. A. Filippov, *American Mathematical Society Translations* **42** series 2 (1964) 199-231.
53. J. Guckenheimer, personal communication (1989).
54. P. Singh and D. D. Joseph, *Physics Letters A* **135** (1989) 247-253.

55. B. Feeny and F. C. Moon, *Nonlinear Dynamics* **4** (1993) 25-37.
56. M. J. Feigenbaum, *Journal of Statistical Physics* **19** (1) (1978) 25-52.
57. N. Metropolis, M. L. Stein, and P. R. Stein, *Journal of Combinatorial Theory* **15**(1) (1973) 25-44.
58. H. G. Schuster, *Deterministic Chaos: an Introduction* (VCH Publishers, Weinheim, Germany, 1988).
59. K. Coffman, W. D. McCormick, and H. L. Swinney, *Physical Review Letters* **56**(10) (1986) 999-1002.
60. M. Kennel, R. Brown, H. D. I. Abarbanel, *Physical Review A* **45** (1992) 3403-3411.
61. H. Abarbanel, *Smart Structures, Nonlinear Dynamics, and Control*, edited by A. Guran and D. J. Inman (Prentice Hall, Up Saddle River, 1995) 1-86.
62. F. Takens, *Lecture Notes in Mathematics* **898** (Springer-Verlag, 1980) 366-381.
63. A. M. Fraser, *IEEE Transactions on Information Theory* **35** (1989) 245-262.

Dynamics with Friction: Modeling, Analysis and Experiment, pp. 93–136
edited by A. Guran, F. Pfeiffer and K. Popp
Series on Stability, Vibration and Control of Systems Series B: Vol. 7
© World Scientific Publishing Company

DYNAMICS OF HOMOGENEOUS FRICTIONAL SYSTEMS

JOSE INAUDI

Earthquake Engineering Research Center
University of California at Berkeley
Berkeley, California 94720, U.S.A.

and

JAMES KELLY

Earthquake Engineering Research Center
University of California at Berkeley
Berkeley, California 94720, U.S.A.

ABSTRACT

This chapter is concerned with the dynamical response of structures with supplemental friction dampers in which resistance increases proportionally to deformation providing triangular hysteresis loops when loaded cyclically. Although nonlinear, the force-deformation relation of this damper is homogeneous of order one; i.e., if the deformation history is scaled, the damper force is scaled by the same factor. Potential applications of this damper are truss structures, tuned mass damper systems, and retrofit upgrade construction of buildings. The free-vibration response of single-degree-of-freedom and multi-degree-of-freedom structures with friction dampers is studied. It is demonstrated that the period of oscillation and the decay ratio of these structures are independent of the vibration amplitude and that in certain cases, structures with this type of damper show mode shapes in free vibration. Linearization techniques are applied to estimate the response of multi-degree-of-freedom structures with dampers subjected to harmonic and broad-band excitation: the harmonic linearization method and a linearization based on a linear hysteretic element. Both linearizations show excellent accuracy. Experimental data from laboratory model testing of the damper is presented and correlated with analytical predictions to validate the models proposed.

1. Introduction

1.1. Damping in earthquake-resistant design of structures

Significant innovation has taken place during the last two decades in the field of earthquake engineering. Technologies such as base isolation, energy dissipation devices and active control systems have been implemented in new construction and in retrofit of existing building structures[9,30]. These vibration protection systems consist of mechanical devices such as isolators, dampers, or hydraulic actuators, which connected to a structural system enhance its dynamical performance. Isolation systems, for example, provide a structural system with a flexible interface such that when the structure is subjected to seismic excitation, significant deformation occurs in the isolation system and small deformation demands are imposed on the superstructure. As their denomination suggests, energy dissipating devices (EDDs) are mechanical dampers or shock absorbers that provide significant energy absorption under cyclic deformation and are designed to endure cyclic deformation without significant damage. When connected to a structural system, EDDs augment its damping. If appropriately provided, this increase in the energy dissipation capacity of the structure can yield reductions in the deformation demand on structural systems subjected to wind or earthquake excitation.

A great variety of EDDs has been proposed and several materials have been used in those EDDs. These include ductile metals, viscoelastic polymers, viscous fluids, shape-memory alloys, and frictional materials. Dry friction, also referred to as Coulomb friction, is a widely used method for energy dissipation. Among the friction dissipators utilized in building structures we find brake lining pads introduced at the intersection of frame cross-braces[24,25], friction dampers consisting of copper pads in contact with steel casing of the device[1], and slotted bolted connections[8,10]. Friction is also utilized for energy dissipation in isolation systems[30]. These friction dampers provide rectangular hysteresis under cyclic deformation because the contact force between the frictional surfaces is approximately constant during operation of the damper and independent of the deformation of the device. This implies that the device is activated at constant force and no energy dissipation is achieved under low excitation levels.

In a linear dissipating mechanism there is a quadratic dependency of the energy dissipated per cycle on the deformation amplitude and dissipation occurs at all excitation levels. This characteristic can be achieved using friction (a nonlinear mechanism) by combining friction dampers and nonlinear kinematic mechanisms[11] or by the sliding of frictional surfaces with contact forces increasing linearly with deformation[4]. An example of a friction device using variable contact force is the ring spring[29] which comprises stacks of concentric inner and outer rings with interactive tapered surfaces which slide across each other. The work reported herein deals with the modeling and analysis of structures containing another device of this type known as the Energy Dissipating Restraint (EDR). Section 2 is devoted to the description of the mechanics of this damper; section 3 deals with the dynamical response of structures containing EDR dampers; section 4 presents linearization methods suitable for the analysis of structures with supplemental EDR dampers; section 5 is devoted to the validation of the proposed mathematical models using data obtained in laboratory experiments of a steel-frame model and a tuned mass damper subjected to simulated ground motion on a shake table; and section 6 presents the main conclusions of this work and suggests directions for further research.

2. The Energy Dissipating Restraint

The Energy Dissipating Restraint is a mechanical damper based on a friction mechanism in which the contact force between the friction surfaces of the device increases linearly with the deformation of the device. Recently proposed by Richter et al.[27] and studied by Nims et al.[22] and Inaudi et al.[14,17], this passive device can be used to increase the energy dissipation in structures to reduce wind, earthquake, and man-induced vibrations. The EDR has been patented in the United States and other countries by Fluor Daniel, Inc.[23].

Figure 1 shows an external view and a cross section of the EDR. The device consists of a steel cylinder, a steel shaft, bronze friction wedges, steel compression wedges, a helical spring, and internal stops. The deformation of the device $\Delta(t)$ is defined as the relative motion of the shaft with respect to the cylinder. The resistance of the device is provided by frictional forces that develop between the bronze frictional wedges and the internal cylinder wall. These frictional forces are functions of the friction coefficient between bronze and steel and the normal force applied by the bronze friction wedges on the cylinder wall. This normal force is, in turn, a result of the compression force in the spring transmitted through the steel compression wedges to the friction wedges. The spring of the EDR can be pre-stressed in the undeformed configuration of the EDR ($\Delta = 0$); the compression force in the spring in the undeformed configuration will be called preload.

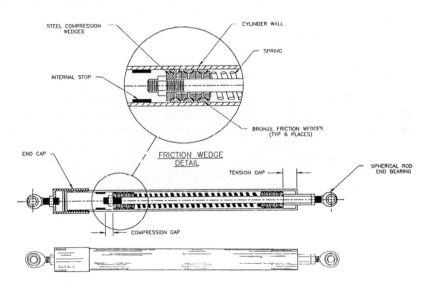

Figure 1. External and internal views of the Energy Dissipating Restraint.

The EDR can be configured to provide several hysteretic behaviors by changing the preload of the spring and the lengths of the tension and compression gaps (Fig. 1). When the preload of the spring is non-zero and the tension and compression gaps are larger than the deformation of the device, the device behaves like a conventional Coulomb friction damper because the compression force in the spring is constant and, hence, the normal force between the bronze wedges and the cylinder wall is constant, leading to a resistance force independent of the deformation amplitude and a rectangular hysteresis loop under cyclic deformation (Fig. 2a). A different behavior is achieved when the spring is pre-stressed and the tension and compression gaps have zero length. In this case, the axial force in the spring varies proportionally to the deformation, leading to flag-shaped hysteresis loops under cyclic loading (Fig. 2b). When the spring has no preload and the lengths of both compression and tension gaps are zero, the spring force increases proportionally to the deformation, leading to triangular hysteresis loops under cyclic deformation (Fig. 2c). Naturally, other combinations are possible. For example, Fig. 2d illustrates the hysteresis loops of the EDR configured with non-zero preload and non-zero lengths for the tension and compression gaps. In this case, the axial force in the spring is constant for deformations smaller than the gap length, but increases proportionally to the difference between the deformation and the gap length for deformations in tension or compression larger than the corresponding gap length. In this chapter, the configuration of the EDR providing triangular hysteresis loops in cyclic deformation (Fig. 2c) is studied.

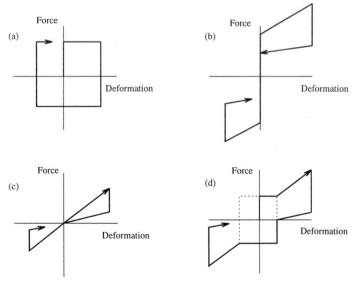

Figure 2. Hysteresis loops of the EDR in four different configurations.

2.1. Mechanics of the EDR damper

Consider a single friction wedge as shown in Fig. 3. The friction force F_f developed at the interface is

$$F_f = \mu \, F_N \tag{1}$$

where μ is the coefficient of friction of bronze and steel, and F_N is the normal contact force between the bronze wedge and the steel cylinder. F_N can be related to the vertical components of the contact forces between the bronze friction wedges and the steel compression wedges which, in turn, are functions of the forces applied on the wedge in the longitudinal and radial directions,

$$F_N = F_{1R} + F_{2R} = F_{1A} \, \cot \alpha + F_{2A} \, \cot \alpha \tag{2}$$

where $\cot \alpha = 1/\tan \alpha$, α is the angle of the tapered friction wedge, F_{1A} is the axial load on right side of the wedge in the direction of motion, F_{2A} is the axial load on left side of the wedge in the direction of motion, F_{1R} is the radial load on the right side of the wedge, and F_{2R} is the radial load on the left side of the wedge. Upon loading, i.e. when $\Delta \dot{\Delta} > 0$,

$$F_{2A} = F_{1A} + F_f \tag{3}$$

while upon unloading, i.e. when $\Delta \dot{\Delta} < 0$,

$$F_{2A} = F_{1A} - F_f \tag{4}$$

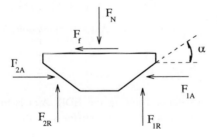

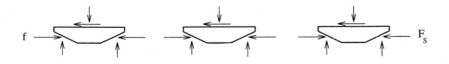

Figure 3. Forces acting on a friction wedge during damper operation.

Replacing Eqs. (1) and (2) into Eq. (3), we obtain

$$F_{2A} = F_{1A} \frac{1 + \mu \ cot \ \alpha}{1 - \mu \ cot \ \alpha} \qquad \Delta \ \dot{\Delta} > 0 \tag{5}$$

for the loading condition; while from Eqs. (1) , (2) and (4) we obtain

$$F_{2A} = F_{1A} \frac{1 - \mu \ cot \ \alpha}{1 + \mu \ cot \ \alpha} \qquad \Delta \ \dot{\Delta} < 0 \tag{6}$$

for the unloading condition. In the case of a single wedge, F_{1A} is the force provided by the spring which in the case on no preload is

$$F_s = K_s \ \Delta \tag{7}$$

where K_s is the stiffness of the spring, and F_{2A} is the resistance force f of the device. Therefore, for the device in loading we can write

$$f = \frac{1 + \mu \ cot \ \alpha}{1 - \mu \ cot \ \alpha} K_s \ \Delta \qquad \Delta \ \dot{\Delta} > 0 \tag{8}$$

while in unloading

$$f = \frac{1 - \mu \ cot \ \alpha}{1 + \mu \ cot \ \alpha} K_s \ \Delta \qquad \Delta \ \dot{\Delta} < 0 \tag{9}$$

Equations (8) and (9) define the force-deformation relation of the EDR for a single friction wedge. If N_w friction wedges are connected in tandem (see bottom figure in Fig. 3), it can be readily shown that

$$f(t) = \begin{cases} v \ K_s \ \Delta(t) & \Delta \ \dot{\Delta} > 0 \quad (loading) \\ \dfrac{K_s}{v} \ \Delta(t) & \Delta \ \dot{\Delta} < 0 \quad (unloading) \end{cases}$$

$$\frac{K_s}{v} \ \Delta(t) \le f(t) \le v \ K_s \ \Delta(t) \qquad \dot{\Delta} = 0 \quad (stick-friction \ regime) \tag{10}$$

where $f(t)$ represents the resistance force in the EDR, $\Delta(t)$ is the deformation of the damper, $\dot{\Delta}(t)$ is the deformation rate, K_s is the stiffness of the spring contained in the damper, and $v \ (> 1)$ is a dimensionless parameter that depends on the number and geometry of the frictional wedges of the damper and the coefficient of friction between the frictional wedges and the internal cylinder wall of the damper

$$v = (\frac{1 + \mu \ cot \ \alpha}{1 - \mu \ cot \ \alpha})^{N_w} \tag{11}$$

As shown in Eq. (10) and in Fig. 4, the EDR is a variable-stiffness device with loading stiffness $v \ K_s$, and unloading stiffness K_s/v. When $\dot{\Delta} = 0$ and $\Delta \ne 0$, the damper shows a stick-friction regime similar to that of a conventional Coulomb friction damper. This means that when $\dot{\Delta} = 0$, the device can withstand any load in the interval $[K_s \Delta/v \ \ v K_s \Delta]$ without increase or decrease in its deformation.

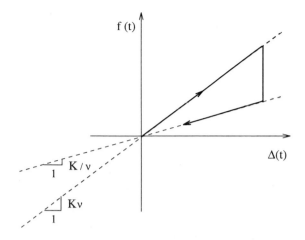

Figure 4. Force-deformation relation for an EDR damper.

In the EDR model defined in Eq. (10), the force is a homogeneous function of order one of the deformation because $f[\alpha \Delta(t)] = \alpha f[\Delta(t)]$. Figure 5 shows force histories for a sinusoidal deformation cycle of amplitude A and frequency 2π *rad/s* for several values of v. The figure shows that for $v = 1$, the normalized force $f(t)/AK_s$ is a sine function, and the EDR behaves like a linear spring. For $v > 1$, the force signal shows jumps at the times at which unloading starts. These jumps are larger for larger v. The force of the EDR subjected to sinusoidal deformation is periodic with the same period of the deformation signal; however, it shows discontinuities which are indicative of the presence of higher harmonics in the force signal and of the nonlinearity of this force-deformation relation. These higher harmonics can be seen in Fig. 6, which shows the amplitude of the Fourier-series coefficients for the normalized force signal for $v = 1, 2, 5, 10$. While for $v = 1$, only the first harmonic is nonzero, for $v > 1$, all odd harmonics are non-zero. From this analysis it can be seen that the behavior of an EDR deviates from that of a linear system as v increases.

In Fig. 4 we note that the hysteresis loops of an EDR damper show a well-defined apparent stiffness equal to the average of the loading and unloading stiffnesses $K_m = K_s(v + 1/v)/2$. The energy dissipation in a sinusoidal cycle of amplitude A can be easily computed as $W_D = K_s A^2 (v - 1/v)$. The normalized apparent stiffness K_m/K_s and the normalized energy dissipation $W_D/(K_s A^2)$ increase with v. This means that for a given spring stiffness K_s, an increase in v that can be achieved by an increase in the number of wedges or a change in the geometry of the wedges, provides a stiffer EDR damper with larger energy dissipation capacity.

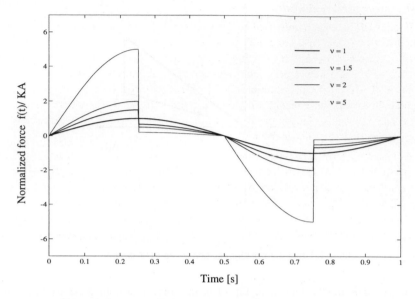

Figure 5. Force in an EDR damper subjected to sinusoidal deformation.

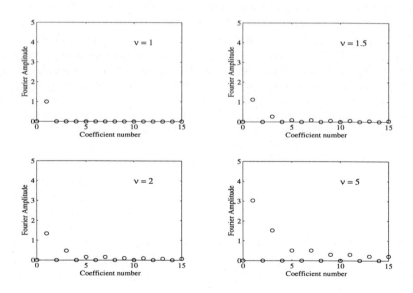

Figure 6. Amplitude of the Fourier-series coefficients of the EDR force.

2.2. The linear-friction element

If an EDR damper configured to provide the behavior described in Eq. (10) is attached to two points of a structure by means of some elastic connection device such as braces or rods (Fig. 5), the relation between the relative motion of the points of attachment and the force applied on those points by the mechanical device is piece-wise linear. Under cyclic deformation, the hysteresis loops of the assembly will still be triangular; however, the element will show a finite stiffness in the transition between loading and unloading due to the finite stiffness of the connector. This constitutive relation is referred to as linear-friction herein.

A hysteresis loop of a linear-friction element is shown in Figure 8. In this figure, K_1 represents the stiffness in loading, K_2 is the stiffness in unloading, and K_3 is the tangent stiffness of the transition between loading and unloading. When unloading occurs ($\Delta \dot{\Delta} < 0$) at a positive deformation Δ_l, the tangent stiffness of the element is finite and given by K_3. If the deformation reaches Δ_u, the tangent stiffness changes to K_2. The tangent stiffness of the element is K_3 while the deformation Δ satisfies

$$\Delta_l(t) < \Delta(t) < \Delta_u(t) \tag{12}$$

where Δ_l is the previous unloading deformation, and Δ_u satisfies

$$\frac{\Delta_u}{\Delta_l} = \frac{K_3 - K_1}{K_3 - K_2} \tag{13}$$

Similarly, when loading starts at a positive deformation Δ_u, the tangent stiffness of the element is K_3 provided the deformation satisfies Eq. (12) with Δ_l satisfying Eq. (13). If the deformation reaches Δ_l in the loading process, the tangent stiffness changes to K_1. Since the device is passive, it can only store and dissipate energy; this implies that the loading stiffness K_1 must be greater than the unloading stiffness K_2. In fact, if an elastic device with stiffness K_b is connected in series with an EDR damper, the parameters K_1, K_2 and K_3 are given by:

$$K_1 = \frac{v\, K_s\, K_b}{v\, K_s + K_b}\, , \quad K_2 = \frac{K_s\, K_b}{K_s + v\, K_b}\, , \quad K_3 = K_b \tag{14}$$

Since $v > 1$ (see Eq. (11)), $K_2 < K_1 < K_3$ for all $K_s > 0$ and $K_b > 0$.

In order to characterize the force-deformation relation of a linear-friction element, it is convenient to define the loading function $l(t) = sgn\,(\Delta(t)\dot{\Delta}(t))$ where $sgn\,(x)$ is the signum function; $sgn\,(x) = 1$ if $x > 0$, $sgn\,(x) = -1$ if $x < 0$, and $sgn\,(x) = 0$ if $x = 0$; and the loading-unloading memory operator $H\,[\Delta(t)]$ which yields, for any time t, the value of the deformation signal $\Delta(t)$ at the immediately prior instant when loading or unloading starts; for a differentiable signal $\Delta(t)$

$$H\,[\Delta(t)] = \Delta(t-\tau)\, , \quad 0 \le \tau = \min(\tau')\, , \quad \tau' = \left\{ u\, :\, \dot{\Delta}(t-u) = 0 \right\} \tag{15}$$

Figure 7. Truss including an EDR damper.

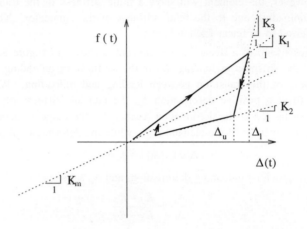

Figure 8. Force-deformation relation of a linear-friction element.

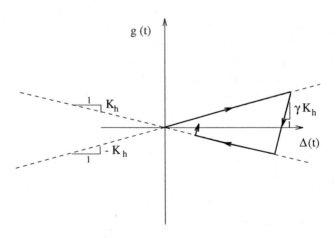

Figure 9. Hysteretic component of the constitutive relation of a linear-friction element.

From the definitions given, we note that $H[\Delta(t)]$ is homogeneous of order one in the deformation; i.e., $H[\alpha\Delta(t)] = \alpha H[\Delta(t)]$. $l(t)$ is invariant under scaling of the deformation signal. The output $z(t)$ of the operator $H[\Delta(t)]$ is a piece-wise constant function of time with jumps at the instants of reversal in the direction of the deformation or instants of zero deformation rate. The functions $\Delta_u(t)$ and $\Delta_l(t)$ used in Eqs. (12) and (13) can be related to $z(t)$ as follows. During loading phases $(l(t) = 1)$,

$$\Delta_u(t) = z(t) , \quad \Delta_l(t) = \Delta_u(t)\frac{K_3 - K_2}{K_3 - K_1} \tag{16}$$

while during unloading phases $(l(t) = -1)$,

$$\Delta_l(t) = z(t) , \quad \Delta_u(t) = \Delta_l(t)\frac{K_3 - K_1}{K_3 - K_2} \tag{17}$$

Using the variable $z(t) = H[\Delta(t)]$, the force in a linear-friction element can be expressed as

$$f(t) = \begin{cases} K_1\,\Delta(t) & l(t) = 1, \ |\Delta(t)| > |z(t)\frac{K_3 - K_2}{K_3 - K_1}| \\[2mm] K_1 z(t)\frac{K_3-K_2}{K_3-K_1} - K_3[z(t)\frac{K_3-K_2}{K_3-K_1} - \Delta(t)] & l(t) = 1, \ |z(t)| < |\Delta(t)| < |z(t)\frac{K_3-K_2}{K_3-K_1}| \\[2mm] K_2\,\Delta(t) & l(t) = -1, \ |\Delta(t)| < |z(t)\frac{K_3 - K_1}{K_3 - K_2}| \\[2mm] K_1 z(t) - K_3[z(t) - \Delta(t)] & l(t) = -1, \ |z(t)\frac{K_3 - K_1}{K_3 - K_2}| < |\Delta(t)| < |z(t)| \end{cases}$$

$$z(t) = H[\Delta(t)], \quad l(t) = sgn(\Delta\,\dot{\Delta}) \tag{18}$$

From Eq. (18) it can be readily shown that the linear-friction element is homogeneous of order one in the deformation Δ. It is also simple to show that Eq. (18) fails to satisfy the additivity (superposition) principle of linear systems. It is worth mentioning that, for a continuous and piece-wise differentiable deformation signal $\Delta(t)$, the force in this model is continuous and piece-wise differentiable. For this reason this model leads to well-posed mathematical problems in static and dynamical analysis of structures containing EDR dampers.

It is convenient to write Eq. (18) as

$$f(t) = K_m\,\Delta(t) + g(\Delta(t), l(t), z(t); K_h, \gamma) \tag{19}$$

where

$$K_m = \frac{K_1 + K_2}{2} \tag{20}$$

and the hysteretic component

$$
g(t) = \begin{cases} K_h \, \Delta(t) & l(t) = 1, \;\; |\Delta(t)| > |z(t)\dfrac{\gamma+1}{\gamma-1}| \\[2ex] K_h \, z(t)\dfrac{\gamma+1}{\gamma-1} - \gamma K_h \, [z(t)\dfrac{\gamma+1}{\gamma-1} - \Delta(t)] & l(t) = 1, \;\; |z(t)| < |\Delta(t)| < |z(t)\dfrac{\gamma+1}{\gamma-1}| \\[2ex] -K_h \, \Delta(t) & l(t) = -1, \;\; |\Delta(t)| < |z(t)\dfrac{\gamma-1}{\gamma+1}| \\[2ex] K_h \, z(t) - \gamma K_h \, [z(t) - \Delta(t)] & l(t) = -1, \;\; |z(t)\dfrac{\gamma-1}{\gamma+1}| < |\Delta(t)| < |z(t)| \end{cases}
$$

$$
z(t) = H[\Delta(t)], \quad l(t) = sgn(\Delta \, \dot{\Delta}) \tag{21}
$$

where

$$
K_h = \frac{K_1 - K_2}{2} \tag{22}
$$

and

$$
\gamma = \frac{K_3 - K_m}{K_h} \tag{23}
$$

The force-deformation relation $g(\Delta, l, z)$ is illustrated in Fig. 9. Because $K_2 < K_1 < K_3$, $1 < \gamma < \infty$; $\gamma \to 1$ when $K_2 \to K_1$; and $\gamma \to \infty$ when $K_3 \to \infty$ for finite K_1 and K_2.

3. Dynamics of structures with EDR dampers

The response of structures with linear-friction elements is studied in this section. It is shown that the period of oscillation and the decay ratio of single-degree-of-freedom (SDOF) systems with EDR dampers are independent of the vibration amplitude. The existence of modes of vibration in multi-degree-of-freedom (MDOF) systems is demonstrated and linearization techniques are applied to estimate the response of multi-degree-of-freedom structures with linear-friction elements subjected to harmonic and broad-band excitation.

3.1. Free vibration of SDOF systems with linear-friction elements

Consider a SDOF system with a linear-friction element described by

$$
\ddot{y}(t) + \omega^2 \, y(t) + \frac{1}{m} \, g(y(t), l(t), z(t); K_h, \gamma) = 0 \tag{24}
$$

with initial conditions $y(0) = Y_n$, $\dot{y}(0) = 0$, and $z(0) = y_o$. In Eq. (24), $\omega^2 = (K_{st} + K_m)/m$, where K_{st} and m are the stiffness and mass of the structure without the EDR damper and K_m is given in Eq. (20). The function $g(t)/m$ takes three different expressions depending upon the variable $z(t) = H[y(t)]$, the deformation $y(t)$, and the deformation rate $\dot{y}(t)$; these expressions are

$$\frac{g(t)}{m} = \zeta\omega^2 y(t) \quad (Loading)$$

$$\frac{g(t)}{m} = -\zeta\omega^2 y(t) \quad (Unloading)$$

$$\frac{g(t)}{m} = \zeta\omega^2 y_l(t) - \gamma\zeta\omega^2(y_l(t) - y(t)) \quad (Transition) \tag{25}$$

where $y_l(t)$ is the unloading deformation, $y_l(t) = z(t)\frac{\gamma-1}{\gamma+1}$ if $l(t) = 1$ and $y_l(t) = z(t)$ if $l(t) = -1$; and $\zeta = K_h/(\omega^2 m)$.

Defining the dimensionless variable $\theta = \omega t$, Eq. (24) can be written as

$$y''(\theta) + y(\theta) + u(y(\theta), l(\theta), z(\theta); \zeta, \gamma) = 0 \tag{26}$$

where $u(t) = \frac{g(t)}{m\omega^2}$, $y' = dy/d\theta$ and $z(\theta) = H[y(\theta)]$. In the variable θ, Eq. (24) can be written as

$$y''(\theta) + (1 + \zeta)\, y(\theta) = 0 \quad (Loading)$$

$$y''(\theta) + (1 - \zeta)\, y(\theta) = 0 \quad (Unloading)$$

$$y''(\theta) + (1 + \gamma\zeta)\, y(\theta) = y_l(\theta)\, \zeta\, (\gamma - 1) \quad (Transition) \tag{27}$$

Equation (27) can be solved exactly for specific initial conditions by solving the corresponding linear system in each region of the state space and ensuring the corresponding continuity conditions.

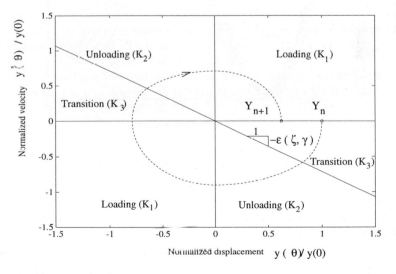

Figure 10. Free vibration of a SDOF in the phase plane ($\zeta = 0.3, \gamma = 10$).

In free vibration, this system presents fixed switching lines in the state space defined by $y(\theta)$ and $y'(\theta)$. At those switching lines, the tangent stiffness of the mechanical device changes. As shown in Fig. 10, the switching lines are given by

$$y(\theta) = 0 , \qquad y'(\theta) = 0 \qquad \text{and} \qquad y'(\theta) = -\varepsilon\, y(\theta) \tag{28}$$

where the constant ε can be computed as

$$\varepsilon = \frac{\gamma+1}{\gamma-1}\ [\ 2\zeta\ \frac{(\gamma-1)^2}{\gamma+1} + 1 + \gamma\,\zeta - 2\,\zeta\,(\gamma-1) - (1+\gamma\,\zeta)\ (\frac{\gamma-1}{\gamma+1})^2\ ]^{\frac{1}{2}} \tag{29}$$

Some details on this derivation can be found in Appendix A. The effect of γ on the free vibration response of a linear-friction SDOF system is illustrated in Fig. 11 for $\zeta = 0.30$ and $\gamma = 10, 20, 50$ and ∞. Small values of γ yield a reduction of the period (stiffening effect) and a reduction of the effective damping.

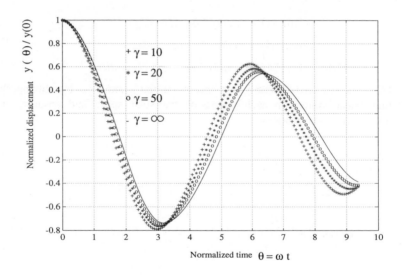

Figure 11. Free vibration of a SDOF with a linear-friction element.

In the case of $\gamma \to \infty$, the transition phase disappears in free vibration. This implies that for the purpose of free vibration analysis, the differential equation of the linear-friction SDOF oscillator with $\gamma \to \infty$, can be reduced to

$$y''(\theta) + y(\theta) + \zeta\, y(\theta)\ sgn\,(y(\theta)y'(\theta)) = 0 \tag{30}$$

The lines $y(\theta) = 0$ and $y'(\theta) = 0$ of the state space constitute the switching lines between two different linear systems with dimensionless frequencies $\sqrt{1+\zeta}$ and $\sqrt{1-\zeta}$.

This equation is the so-called Reid oscillator, which shows a periodic response with a constant decay ratio between consecutive peaks[4]. The dimensionless period of the Reid oscillator, T_θ, is independent of the amplitude of oscillation and given by

$$T_\theta(\zeta) = T \; \omega = \pi \left(\frac{1}{\sqrt{1-\zeta}} + \frac{1}{\sqrt{1+\zeta}} \right) \tag{31}$$

where T represents the period of the system in units of time. The ratio of peaks Y_i between two successive cycles, n and $n+1$, is constant

$$\frac{Y_{n+1}}{Y_n} = \frac{1-\zeta}{1+\zeta} \tag{32}$$

In the case of finite γ, the nonlinear system described in Eq. (25) also exhibits a vibration period, $T_\theta(\zeta, \gamma)$, independent of the amplitude of oscillation

$$T_\theta(\zeta, \gamma) = \pi \left(\frac{1}{\sqrt{1-\zeta}} + \frac{1}{\sqrt{1+\zeta}} \right) - \frac{2}{\sqrt{1-\zeta}} \tan^{-1} \left(\frac{2\sqrt{\gamma+\zeta}}{\sqrt{1-\zeta}\,(\gamma-1)} \right) +$$

$$+ \frac{2}{\sqrt{1+\gamma\zeta}} \cos^{-1} \left(\frac{(\gamma-1)\,(1-\zeta)}{(1+\gamma)\,(1+\zeta)} \right) \tag{33}$$

Naturally, the normalized period is a function of γ and ζ. Figure 12 shows the variation of the period $T_\theta(\zeta, \gamma)$ with γ for different values of ζ. The results are normalized by the period of the system with $\gamma = \infty$, $T_\theta(\zeta)$ (Eq. (31)), and given for $\zeta = 0.2, ,0.5, 0.6$ and 0.8.

Like the Reid oscillator, the linear-friction oscillator (Eq. (25)) shows a constant decay ratio between peaks in free vibration

$$\frac{Y_{n+1}}{Y_n} = \frac{1-\zeta}{1+\zeta} \left(\frac{\gamma-1}{\gamma+1} \right)^2 \left(1 + 4 \frac{\gamma+\zeta}{(1-\zeta)\,(\gamma-1)^2} \right) \tag{34}$$

To analyze the effect of γ on the decay ratio of the oscillator, an apparent damping ratio is computed based on the well-known result of linear viscously-damped systems

$$\frac{Y_{n+1}}{Y_n} = e^{-\frac{2\pi\xi}{\sqrt{1-\xi^2}}} \tag{35}$$

The apparent damping ratio $\xi(\gamma, \zeta)$ of this nonlinear system can be computed from Eq. (35) as

$$\xi(\gamma, \zeta) = \frac{-\dfrac{1}{2\pi} \ln(h\,(\gamma, \zeta))}{\left[1 + \dfrac{1}{4\pi^2} \ln^2(h\,(\gamma, \zeta)) \right]^{\frac{1}{2}}} \tag{36}$$

where $h\,(\gamma, \zeta) = Y_{n+1}/Y_n$ is given in Eq. (34). Figure 13 shows the apparent damping

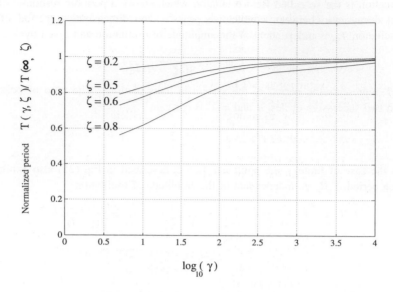

Figure 12.   Effect of γ on the period of oscillation.

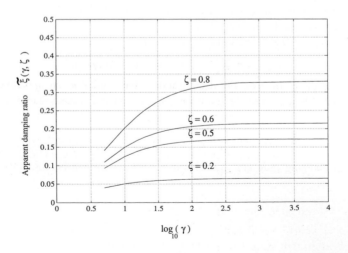

Figure 13.   Effect of γ on the apparent damping ratio of the system.

ratio $\xi(\gamma, \zeta)$ as a function of γ for $\zeta = 0.2, 0.5, 0.6$ and 0.8. The effect of a reduction in γ is to decrease the effective damping ratio. For $\gamma > 100$, the effect of memory on the equivalent damping ratio is negligible and ξ could be taken, without significant error, as that corresponding to $\gamma \to \infty$ (which can be computed using Eqs. (32) and (36)).

As shown in this section, SDOF structures with linear-friction elements exhibit two characteristics of time-invariant linear structures: the vibration period and the logarithmic decrement are independent of the amplitude of oscillation. In fact, these characteristics are shared by all homogeneous SDOF oscillators[12].

3.2. MDOF structures with EDR elements

Consider an N-degree-of-freedom system containing linear-friction dampers. The system can be described by the following differential equation

$$\mathbf{M}\,\ddot{\mathbf{y}}(t) + \mathbf{K}\,\mathbf{y}(t) + \sum_{i=1}^{N_e} \mathbf{L}_i^T f_i(t) = \mathbf{L}_w\,\mathbf{w}(t), \qquad \mathbf{y}(0) = \mathbf{y}_o, \qquad \dot{\mathbf{y}}(0) = \dot{\mathbf{y}}_o \tag{37}$$

where $\mathbf{M}$ and $\mathbf{K}$ represent the positive-definite mass and stiffness matrices, respectively; $\mathbf{L}_w\,\mathbf{w}(t)$ represents the excitation, $f_i(t)$ is the force in the i-th damper, $\mathbf{L}_i^T$ is the appropriate force transformation, and N_e is the number of dampers. The force in the i-th EDR device can be expressed as

$$f_i(t) = K_m^{(i)}\,\Delta_i(t) + g(\Delta_i(t), l_i(t), z_i(t); K_h^{(i)}, \gamma_i) \tag{38}$$

where $K_m^{(i)}$, $K_h^{(i)}$ and γ_i are the parameters of the i-th EDR defined in Eqs. (20), (22), and (23); $z_i(t) = H[\Delta_i(t)]$, $l_i(t) = sgn(\Delta_i \dot{\Delta}_i)$, the element deformation Δ_i in the i-th element is given by

$$\Delta_i(t) = \mathbf{L}_i\,\mathbf{y}(t), \tag{39}$$

and the function $g(t)$ has been defined in Eq. (21).

By defining $\tilde{\mathbf{K}} = \mathbf{K} + \sum_{i=1}^{N_e} \mathbf{L}_i^T K_m^{(i)} \mathbf{L}_i$, Eq. (37) can be written as

$$\mathbf{M}\,\ddot{\mathbf{y}}(t) + \tilde{\mathbf{K}}\,\mathbf{y}(t) + \sum_{i=1}^{N_e} \mathbf{L}_i^T g(\Delta_i(t), l_i(t), z_i(t); K_h^{(i)}, \gamma_i) = \mathbf{L}_w\,\mathbf{w}(t) \tag{40}$$

Since the linear-friction element is homogeneous of order one, the system described in Eq. (40) is also homogeneous; if $\mathbf{y}(t)$ is the solution of Eq. (40) for initial conditions $\mathbf{y}_o$ and $\dot{\mathbf{y}}_o$ and excitation $\mathbf{w}(t)$, $\alpha\,\mathbf{y}(t)$ is the solution for initial conditions $\alpha\,\mathbf{y}_o$ and $\alpha\,\dot{\mathbf{y}}_o$ and excitation $\alpha\mathbf{w}(t)$.

Mode shapes in free vibration of MDOF structures

Structural systems with supplemental EDR dampers with parameters $K_m^{(i)}$, $K_h^{(i)}$ and γ_i such as that described in Eq. (40) can exhibit mode shapes in free vibration when the initial condition is parallel to an eigenvector satisfying

$$\omega_l^2 \, \mathbf{M} \, \phi_l = \tilde{\mathbf{K}} \, \phi_l \quad l = 1, 2, ..., N \tag{41}$$

if the matrix $\sum\limits_{i=1}^{N_e} \mathbf{L}_i^T \, K_h^{(i)} \, \mathbf{L}_i$ is classical in the sense that

$$\phi_j^T \sum_{i=1}^{N_e} \mathbf{L}_i^T \, K_h^{(i)} \, \mathbf{L}_i \, \phi_l = 0, \quad l \neq j \tag{42}$$

and all γ_i are equal

$$\gamma_i = \gamma, \quad i = 1,..., N_e. \tag{43}$$

To demonstrate this statement, consider the hysteretic component of the resistance forces provided by EDR dampers in a structural system for the structure oscillating in a particular mode

$$\mathbf{F}(t) = \sum_{i=1}^{N_e} \mathbf{L}_i^T \, g_i(t) \tag{44}$$

where $\mathbf{F}(t)$ is the vector of forces on the degrees of freedom of the model and $g_i(t)$ is the hysteretic force in the i-th linear-friction element.

Let the displacements of the structure be parallel to a particular direction ϕ_l for all t,

$$y(t) = \phi_l \, q_l(t) \tag{45}$$

where q_l is a modal coordinate. Then,

$$\Delta_i(t) = \mathbf{L}_i \, \phi_l \, q_l(t) \tag{46}$$

and from Eq. (21), the i-th element force is

$$g_i(t) = K_h^{(i)} \, \mathbf{L}_i \, \phi_l \, q_l(t) \tag{47}$$

during loading phases,

$$g_i(t) = -K_h^{(i)} \, \mathbf{L}_i \, \phi_l \, q_l(t) \tag{48}$$

during unloading phases, and

$$g_i(t) = K_h^{(i)} \, \mathbf{L}_i \, \phi_l \, (\gamma \, q_l(t) + H[q_l(t)] \, (1-\gamma)) \tag{49}$$

during transition phases. To obtain Eq. (49) we have used the assumption given in Eq. (43). This implies that

$$\mathbf{F}(t) = \sum_{i=1}^{N_e} \mathbf{L}_i^T \, K_h^{(i)} \, \mathbf{L}_i \, \phi_l \, q_{l(t)} \tag{50}$$

during loading phases,

$$\mathbf{F}(t) = - \sum_{i=1}^{N_e} \mathbf{L}_i^T \, K_h^{(i)} \, \mathbf{L}_i \, \phi_l \, q_l(t) \tag{51}$$

during unloading phases, and

$$\mathbf{F}(t) = \sum_{i=1}^{N_e} \mathbf{L}_i^T K_h^{(i)} \mathbf{L}_i \, \boldsymbol{\phi}_l \, (\gamma \, q_l(t) + H[q_l(t)] \, (1-\gamma)) \tag{52}$$

during transition phases. As Eqs. (50) through (52) show, $\mathbf{F}(t)$ does not change direction when the structure oscillates in a particular mode shape. Furthermore, from Eqs. (50) through (52) and the assumption given in Eq. (41), we conclude that

$$\boldsymbol{\phi}_j^T \, \mathbf{F}(t) = 0 \quad j \neq l \tag{53}$$

This means that no contribution of modes other than l should be expected in free vibration if the initial condition is parallel to $\boldsymbol{\phi}_l$. Therefore, we have proven that if Eqs. (42) and (43) are satisfied and $\mathbf{y}(0) = c \, \boldsymbol{\phi}_l$, $\mathbf{y}(t)$ stays in the same direction showing a modal response (Eq. (45)). Furthermore, the modal coordinate $q_l(t)$ satisfies the differential equation of a SDOF oscillator as described in Eq. (24) with parameters γ,

$$\omega = \omega_l \tag{54}$$

where ω_l is given in Eq. (41), and

$$\zeta = \frac{\boldsymbol{\phi}_l^T \displaystyle\sum_{i=1}^{N_e} \mathbf{L}_i^T K_h^{(i)} \mathbf{L}_i \, \boldsymbol{\phi}_l}{\boldsymbol{\phi}_l^T \, \bar{\mathbf{K}} \, \boldsymbol{\phi}_l} \tag{55}$$

and initial condition $q_l(0) = c$.

4. Linearization techniques for the analysis of structures with EDR dampers

The forced vibration of MDOF structures with classical linear-friction damping can be reduced to the forced-vibration of a SDOF structure when the forcing $\mathbf{L}_w \mathbf{w}(t)$ is orthogonal to all but a single mode shape $\boldsymbol{\phi}_l$. In this very particular case, the solution will stay parallel to $\boldsymbol{\phi}_l$ and can be reduced to the solution of the forced vibration of a SDOF linear-friction oscillator. This unusual feature of nonlinear systems has been studied in detail for a particular class of MDOF structures with Reid elements (linear-friction elements with $\gamma = \infty$) subjected to periodic excitation[5].

Generally speaking, the computation of the response of MDOF structures with linear-friction devices subjected to deterministic or random loading requires numerical integration. The use of response-spectrum techniques in the preliminary design of structures subjected to dynamical loading is common practice in structural engineering. The earthquake-resistant design of structures is typically based on a linear response spectrum. For this reason, obtaining a linear system of equations which approximates Eq. (40) for a given type of excitation is highly desirable. In this section, different linearization techniques are used to approximate the response of this nonlinear system to harmonic and random excitations. These tools are very useful for the preliminary sizing and definition of location of EDR dampers in a structural system.

4.1. Harmonic linearization of a linear-friction element

The harmonic linearization technique, often called the describing-function method or the method of harmonic balance[18] seeks a linear element that approximates the response of a nonlinear element when the input to the nonlinear element is harmonic. Here, we seek an approximate linear element of the form

$$\tilde{f} = k_e \Delta(t) + c_e \dot{\Delta}(t) \tag{56}$$

to replace the nonlinear hysteretic element

$$f = g(\Delta(t), l(t), z(t); K_h, \gamma) \tag{57}$$

defined in Eq. (21). Letting the deformation $\Delta(t) = A \sin(\bar{\omega}t)$, we seek parameters k_e and c_e such that the following integral of square error is minimized:

$$J(c_e, k_e) = \int_0^{\frac{2\pi}{\bar{\omega}}} [\tilde{f}(\Delta(t), \dot{\Delta}(t)) - g(\Delta(t), l(t), z(t))]^2 \, dt \tag{58}$$

After some algebra (see Appendix B for details), the following expressions are obtained for k_e and c_e

$$k_e = K_h \, \rho_k(\gamma) \tag{59}$$

$$c_e = \frac{2 K_h}{\bar{\omega} \pi} \rho_c(\gamma) \tag{60}$$

where $\rho_k(\gamma)$ and $\rho_c(\gamma)$ are given by

$$\rho_k(\gamma) = \frac{1}{\pi} [\cos^{-1}(\frac{\gamma-1}{\gamma+1}) (1+\gamma) - 2\sqrt{\gamma} \frac{\gamma-1}{\gamma+1}] \tag{61}$$

$$\rho_c(\gamma) = \frac{1}{2} [1 + 4\frac{\gamma-1}{\gamma+1} - \frac{4\gamma^2}{(\gamma+1)^2} + (\frac{\gamma-1}{\gamma+1})^2] \tag{62}$$

The functions $\rho_k(\gamma)$ and $\rho_c(\gamma)$ are plotted in Fig. 14; $\rho_k(\gamma)$ is a monotonically decreasing function of γ, $\rho_k(1) = 1$, and $\rho_k(\infty) = 0$. Conversely, $\rho_c(\gamma)$ is a monotonically increasing function of γ, $\rho_c(1) = 0$, and $\rho_c(\infty) = 1$.

It can be shown that the expression for c_e given in Eq. (60) can also be obtained by matching the dissipation of energy in a cycle of deformation of the equivalent viscous damper to that of the linear-friction element (Eq. (21)), E_h. Therefore, E_h can be written as

$$E_h = 2 A^2 K_h \, \rho_c(\gamma) \tag{63}$$

where A is the deformation amplitude.

A characteristic frequency of the deformation response is required to define c_e in Eq. (60). In the case of narrow-band excitation, an appropriate selection of $\bar{\omega}$ is the forcing frequency. To approximate the steady-state response of a SDOF linear-friction

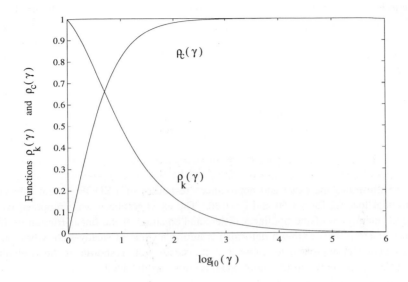

Figure 14. Functions $\rho_k(\gamma)$ and $\rho_c(\gamma)$ obtained using harmonic linearization.

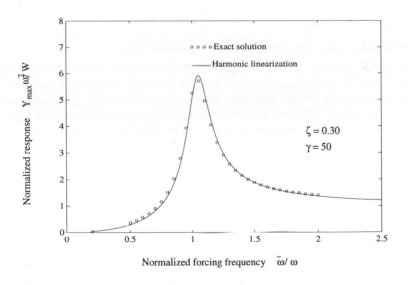

Figure 15. Harmonic response of the linear-friction oscillator and its equivalent linear system.

system to harmonic excitation, the following equivalent linear system can be used

$$\ddot{y}(t) + 2\,\xi_e\,\omega_e\,\dot{y}(t) + \omega_e^2\,y(t) = W\,\sin\beta\omega t \tag{64}$$

where W is the forcing amplitude, the equivalent frequency $\omega_e = \sqrt{\omega^2 + k_e/m}$ is computed using Eq. (59) as

$$\omega_e = \omega\,\sqrt{1 + \zeta\,\rho_k(\gamma)}, \tag{65}$$

and the equivalent damping ratio is computed from $\xi_e = c_e/2m\,\omega_e$ with c_e given by Eq. (60) and taking $\bar{\omega}$ equal to the forcing frequency $\bar{\omega} = \beta\omega$,

$$\xi_e = \frac{\zeta}{\pi\beta}\,\frac{\rho_c(\gamma)}{\sqrt{1 + \zeta\,\rho_k(\gamma)}} \tag{66}$$

Figure 15 illustrates the exact and approximate responses of a SDOF system subjected to sinusoidal loading for $\gamma = 50$ and $\zeta = 0.30$. The exact response was computed using numerical integration of the nonlinear differential equation of the linear-friction oscillator and extracting the maximum deformation during a cycle of steady-state vibration of the oscillator. The approximate response is the steady-state response of the equivalent linear system. As shown in the figure good accuracy is obtained.

In the case of broad-band excitation, the natural frequency of the oscillator is an appropriate choice for $\bar{\omega}$ in Eq. (60); the equivalent damping ratio in this case can be obtained from Eq. (60) taking $\bar{\omega} = \omega_e$ to yield

$$\xi_e = \frac{\zeta}{\pi}\,\frac{\rho_c(\gamma)}{1 + \zeta\,\rho_k(\gamma)} \tag{67}$$

4.2. Linearization using the modal strain energy method

In MDOF systems with well separated frequencies, the Modal Strain Energy method[19,31] can be used in combination with the harmonic linearization technique to estimate the response of the nonlinear system described in Eq. (40) subjected to broad-band excitation. We seek a set of uncoupled differential equations that approximates the response of the nonlinear system. Let us define the classical mode shapes, ϕ_i, and modal frequencies, ω_i, by replacing the nonlinear term of Eq. (40) with the corresponding linear spring elements, $K_e^{(i)}$, and neglecting the equivalent damping elements, $C_e^{(i)}$, then

$$\omega_l^2\,\mathbf{M}\,\phi_i = \Big(\tilde{\mathbf{K}} + \sum_{i=1}^{N_e} \mathbf{L}_i^T\,K_e^{(i)}\,\mathbf{L}_i\Big)\,\phi_l \quad l = 1, 2, \ldots, N \tag{68}$$

$$\phi_i^T\,\mathbf{M}\,\phi_j = \delta_{ij} \tag{69}$$

where $K_e^{(i)}$ is given by Eq. (59) with $K_h = K_h^{(i)}$ and $\gamma = \gamma_i$. By defining the modal coordinates, $\mathbf{q}$, as

$$\mathbf{y}(t) = \Phi\,\mathbf{q}(t) \tag{70}$$

and using the orthogonality property of the modal matrix, $\Phi = [\phi_1 \cdots \phi_N]$, the equivalent linear system can be described by

$$\ddot{q}(t) + \Omega^2 \, q(t) + \Phi^T \sum_{i=1}^{N_e} L_i^T C_e^{(i)} L_i \, \Phi \, \dot{q}(t) = \Phi^T L_w \, w(t) \tag{71}$$

where Ω is a diagonal matrix with terms ω_i^2 in the diagonal. Neglecting the interaction between the modal coordinates due to coupling through the equivalent damping matrix, we obtain the following uncoupled equations of motion

$$\ddot{q}_l(t) + \omega_l^2 \, q_l(t) + \phi_l^T \sum_{i=1}^{N_e} L_i^T C_e^{(i)} L_i \, \phi_l \, \dot{q}_l(t) = \phi_l^T L_w \, w(t) \qquad l = 1, 2,..., N \tag{72}$$

In the case of broad-band excitation, the equivalent damping constant $C_e^{(i)}$ in the l-th equation of Eq. (72) is taken as

$$C_e^{(i)} = \frac{2 \, K_h^{(i)} \, \rho_c(\gamma_i)}{\omega_l \, \pi} \tag{73}$$

Finally, the following modal equations are obtained to approximate the response of the nonlinear system

$$\ddot{q}_l(t) + 2\omega_l \xi_l \, \dot{q}_l(t) + \omega_l^2 \, q_l(t) = \phi_l^T L_w \, w(t) \qquad l = 1, 2,..., N \tag{74}$$

where the equivalent modal damping ratios, ξ_l, are given by

$$\xi_l = \frac{1}{\pi\omega_l^2} \sum_{i=1}^{N_e} \phi_l^T L_i^T \, K_h^{(i)} \, \rho_c(\gamma_i) \, L_i \, \phi_l \, , \qquad l = 1, 2,..., N \tag{75}$$

Figure 16 shows a comparison of the exact and approximate responses of a 2-DOF structure with linear-friction elements subjected to the El Centro earthquake. The parameters of the 2-DOF system considered are

$$M = m \begin{bmatrix} 1 & 0 \\ 0 & 1 \end{bmatrix} \qquad K = m \begin{bmatrix} 200 & -100 \\ -100 & 100 \end{bmatrix} 1/s^2$$

$$L_1^T = \begin{bmatrix} 1 \\ 0 \end{bmatrix} \qquad L_2^T = \begin{bmatrix} 1 \\ -1 \end{bmatrix} \qquad L_w = \begin{bmatrix} -m \\ -m \end{bmatrix} \tag{76}$$

Two linear-friction elements with parameters $K_1/m = 50 \, 1/s^2$, $K_2/m = 1 \, 1/s^2$, and $K_3/m = 300 \, 1/s^2$ are connected to the structure (top figure in Fig. 18). The natural frequencies of the structure without the EDDs are $\omega_1 = 6.18 \, rad/s$ and $\omega_2 = 16.18 \, rad/s$. Using Eqs. (22), (23), (61) and (62) and the assumed parameters K_1, K_2 and K_3, we can compute $\gamma = 11.204$, $\rho_k = 0.474$, $\rho_c = 0.836$. $K_m/m = 25.5 \, 1/s^2$ and the equivalent stiffness of each EDD is $K_e/m = 11.608 \, 1/s^2$. The mode shapes, modal frequencies and modal damping ratios of the equivalent linear system are obtained as

$$\phi_1^T = [0.526 \quad 0.851] \qquad \omega_1 = 7.2 \, rad/s \qquad \xi_1 = 0.048 \tag{77}$$

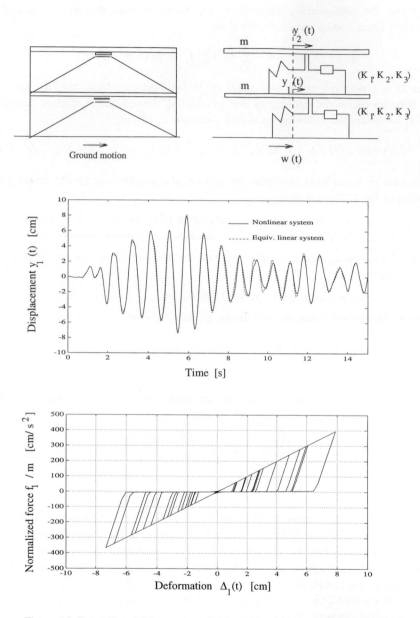

Figure 16. Comparison of the response of a 2-DOF structure with EDR dampers
and its linear model.

$$\phi_2^T = [-0.851 \quad 0.526] \qquad \omega_2 = 18.9 \; rad/s \qquad \xi_2 = 0.048$$

As shown in Fig. 16, excellent accuracy is obtained using the linearization technique in the estimation of the transient response of the nonlinear structure. The bottom figure in Fig. 16 shows the force-deformation relation of the EDR damper installed at the first story of the structure obtained in the simulation of the response of the nonlinear system. It is worth noting that the relative accuracy of this linearization (maximum error in a certain response quantity normalized by the maximum response value) does not depend on the amplitude of the excitation since both this nonlinear system and the equivalent linear system are homogeneous (of order one) in the input excitation. The accuracy of this procedure depends on the level of nonlinearity of the linear-friction elements which increases with $K_h^{(i)}$.

4.3. Statistical linearization of Reid's element

The statistical linearization method is used in this section to estimate the mean square response of structures containing EDR elements with $\gamma = \infty$ subjected to stationary random excitation.

Let us obtain the equivalent damping constant, c_e, of a linear damper that approximates the Reid element $f(t) = K_h \, \Delta \, sgn(\Delta\dot{\Delta})$ by minimizing the mean square error assuming a stationary jointly-Gaussian distribution for the deformation and deformation-rate processes. The coefficient of the equivalent linear element can be computed by solving the following minimization problem:

$$minimize \quad \Lambda[c_e] = E[\varepsilon^2] \tag{78}$$

where the error ε is given by

$$\varepsilon(t) = K_h \Delta(t) \, sgn(\Delta(t)\,\dot{\Delta}(t)) - c_e \dot{\Delta}(t) \tag{79}$$

and the expectation operator $E[.]$ is taken as

$$E[(.)] = \frac{1}{2\pi\sigma_\Delta\sigma_{\dot{\Delta}}} \int_{-\infty}^{\infty} \int_{-\infty}^{\infty} e^{-\frac{x^2}{2\sigma_\Delta}} e^{-\frac{\dot{x}^2}{2\sigma_{\dot{\Delta}}}} (.) \; dx \; d\dot{x} \tag{80}$$

In Eq. (80), σ_Δ and $\sigma_{\dot{\Delta}}$ are the root mean square deformation and deformation rate.

After some algebra, the minimization of the mean square error yields[11]

$$c_e = \frac{2 K_h}{\pi} \frac{\sigma_\Delta}{\sigma_{\dot{\Delta}}} \tag{81}$$

The equivalent damping parameter depends on the response of the structure to which the element is connected through the ratio $\sigma_\Delta/\sigma_{\dot{\Delta}}$. This ratio is a function of the properties of the structural system, those of the EDD, and the characteristics of the stationary input process. It is interesting to note that c_e is independent of the intensity of the excitation process and only dependent on its frequency content if the dampers are connected to a linear structure.

The stationary response of the nonlinear system described in Eq. (40) (with $\gamma_i = \infty$) subjected to random loading can be estimated using the equivalent linear system. The equivalent linear system presents the same mass and stiffness matrices as those of the nonlinear system and the following damping matrix (which has meaning only under a stochastic analysis)

$$\mathbf{C}_e = \sum_{i=1}^{N_e} \mathbf{L}_i^T c_e^{(i)} \mathbf{L}_i \tag{82}$$

where $c_e^{(i)}$ is given by

$$c_e^{(i)} = \frac{2 \, K_h^{(i)}}{\pi} \frac{\sigma_{\Delta_i}}{\sigma_{\dot{\Delta}_i}} \tag{83}$$

where the root mean square deformation, σ_{Δ_i}, and the root mean square deformation rate, $\sigma_{\dot{\Delta}_i}$, are given by

$$\sigma_{\Delta_i} = [\, \mathbf{L}_i \, E[\mathbf{y}\mathbf{y}^T] \, \mathbf{L}_i^T \,]^{\frac{1}{2}} \tag{84}$$

$$\sigma_{\dot{\Delta}_i} = [\, \mathbf{L}_i \, E[\dot{\mathbf{y}}\dot{\mathbf{y}}^T] \, \mathbf{L}_i^T \,]^{\frac{1}{2}}$$

To illustrate the procedure, let us obtain the mean square response of the nonlinear system described by Eq. (40) subjected to a zero-mean white process with autocorrelation function

$$E[\mathbf{w}(t) \, \mathbf{w}(t+\tau)] = \mathbf{W} \, \delta(\tau) \tag{85}$$

The stationary state covariance matrix of the equivalent linear system can be obtained by solving the following matrix equation[28]

$$\mathbf{A}_e \, \mathbf{P} + \mathbf{P} \, \mathbf{A}_e^T = -\mathbf{B}_w \, \mathbf{W} \, \mathbf{B}_w^T \tag{86}$$

where

$$\mathbf{P} = E[\mathbf{x}\mathbf{x}^T] \qquad \mathbf{x}^T = [\mathbf{y}^T \; \dot{\mathbf{y}}^T] \qquad \mathbf{A}_e = \begin{bmatrix} \mathbf{O} & \mathbf{I} \\ -\mathbf{M}^{-1}\tilde{\mathbf{K}} & -\mathbf{M}^{-1}\mathbf{C}_e \end{bmatrix} \qquad \mathbf{B}_w^T = \begin{bmatrix} \mathbf{O} \\ \mathbf{M}^{-1}\mathbf{L}_w \end{bmatrix} \tag{87}$$

Numerical methods are required to solve for the state covariance matrix $\mathbf{P}$ because Eq. (87) is a set of nonlinear algebraic equations. (Equation (87) is nonlinear because $\mathbf{A}_e$ depends on $\mathbf{P}$ through the mean square deformations σ_{Δ_i} and the mean square deformation rates $\sigma_{\dot{\Delta}_i}$ which in turn depend on $\mathbf{P}$.) An iterative technique was used in this study to solve for the stationary state covariance matrix.

The 2-DOF system described previously is used for a numerical example on the accuracy of this linearization method. The parameters of the Reid elements are taken as $K_1/m = 50 \ 1/s^2$, $K_2/m = 1 \ 1/s^2$ and $K_3/m = \infty$. Monte Carlo simulation techniques are used to estimate the response of the nonlinear system subjected to white-noise support acceleration $(E[w(t)w(t+\tau)] = 1000 \ cm^2/s^3)$. The mean square response of the nonlinear system was estimated by averaging the results of 250 numerical simulations. The stationary response of the equivalent linear system was computed using Eqs. (82) through (86). Figure 17 compares the mean square response obtained by Monte Carlo simulation with the stationary response computed using the statistical linearization technique. Very satisfactory accuracy is obtained.

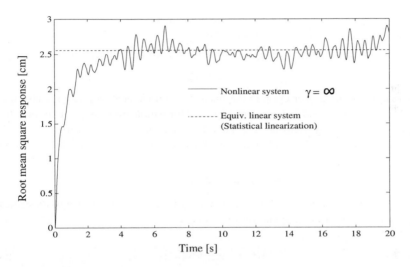

Figure 17. Monte Carlo simulation results and statistical linearization results (Reid's model).

4.4. *Linearization using a linear hysteretic damping model*

An equivalent Kelvin element is typically used to linearize nonlinear elements. In this section, a linearization based on a linear hysteretic model is proposed to estimate the response of the nonlinear system to stationary random excitation.

Many materials show force-deformation relations which are independent of the deformation-rate amplitude. This behavior is usually referred to as hysteretic. According to Bishop[2], linear models for hysteretic damping were proposed prior to 1937. Such denominations as linear hysteretic damping, structural damping, or complex-valued stiffness, have been used in the literature to refer to a linear model of damping in which the energy loss per cycle is independent of the deformation frequency. Interesting discussions on this model can be found in the literature[2,7,21,26].

A linear hysteretic model that shows energy loss per cycle quadratic in the deformation amplitude and independent of the frequency can be given in the frequency domain by

$$F_l(j\varpi) = j \; sgn(\varpi) \; S_h \; \Delta(j\varpi) \tag{88}$$

where $j = \sqrt{-1}$, S_h is a real-valued parameter with units of stiffness, $\Delta(j\varpi)$ is the Fourier transform of the deformation, and $F_l(j\varpi)$ is the Fourier transform of the force of the model. This linear hysteretic element is non-causal[7], i.e., the force anticipates the deformation history in transient vibrations; this naturally violates the principle that in physical systems effect cannot precede cause. For this reason, this model has been mostly used in the context of harmonic excitation.

The energy dissipation E_{lh} in a deformation cycle of amplitude A in this model is

$$E_{lh} = \pi \, A^2 \, S_h \tag{89}$$

The dissipation of energy per cycle in the linear-friction element (Eq. (21)) is also quadratic in the amplitude and independent of the deformation frequency. It follows that a complex-valued spring element is a suitable candidate for the linearization of the linear-friction element. The following equivalent linear model in the frequency domain is proposed

$$\tilde{F}(j\varpi) = (K_e + j \; sgn(\varpi) \; S_e) \; \Delta(j\varpi) \tag{90}$$

where K_e is given by Eq. (59) and S_e is obtained by matching the energy dissipation of both models. From Eqs. (63) and (89) we obtain

$$S_e = \frac{2 \, K_h}{\pi} \, \rho_c(\gamma) \tag{91}$$

The stationary response of the equivalent linear hysteretic system can be computed using frequency-domain techniques for a given power spectral density of the excitation process. The dynamics of the equivalent linear structure can be expressed in the frequency domain as

$$[(j\varpi)^2 \, \mathbf{M} + \tilde{\mathbf{K}} + \sum_{i=1}^{N_e} \mathbf{L}_i^T \, (K_e^{(i)} + j \; sgn(\varpi) \; S_e^{(i)}) \, \mathbf{L}_i \,] \, \mathbf{Y}(j\varpi) = \mathbf{L}_w \, \mathbf{W}(j\varpi) \tag{92}$$

where $K_e^{(i)}$ and $S_e^{(i)}$ are given by Eqs. (59) and (91), respectively, with the parameters of the i-th element. The stationary mean square response to a stationary excitation can be computed as

$$E[\, \mathbf{y} \, \mathbf{y}^T \,] = \int_{-\infty}^{\infty} \mathbf{H}_{wy}(j\varpi) \, \mathbf{S}_w(\varpi) \, \mathbf{H}_{wy}^*(j\varpi) \, d\varpi \tag{93}$$

where $\mathbf{H}_{wy}(j\varpi)$ is the transfer function from $\mathbf{w}$ to $\mathbf{y}$, $\mathbf{S}_w(\varpi)$ is the power spectral density of the excitation, and $()^*$ represents complex transposition.

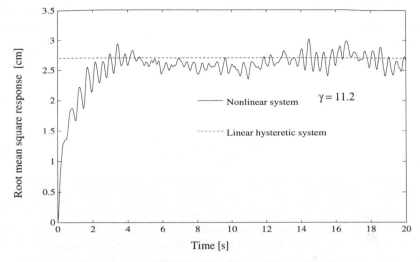

Figure 18. Monte Carlo simulation results and hysteretic linearization results.

A simple numerical example is developed to illustrate the accuracy of this method in the estimation of the mean square response of the nonlinear system containing linear-friction elements. The mean square response of the nonlinear system described by Eq. (40) subjected to support acceleration modeled by a zero-mean white process with autocorrelation function

$$E[w(t)\,w(t+\tau)] = W\,\delta(\tau) \tag{94}$$

is computed using Monte Carlo simulation techniques. The 2-DOF system described previously was subjected to white noise excitation of intensity $W = 1000\ cm^2/s^3$. The parameters of the linear-friction elements are $K_1/m = 50\ 1/s^2$, $K_2/m = 1\ 1/s^2$, and $K_3/m = 300\ 1/s^2$. The stationary root mean square displacement of the nonlinear system was computed by averaging in time the mean square deformation response obtained in a Monte Carlo simulation (250 simulations) between $t = 10\ s$ and $t = 20\ s$, giving

$$\sigma_{y_1} = \sqrt{E[y_1^2]} = 2.66\ cm \tag{95}$$

Using the hysteretic linearization procedure, we can estimate the mean square response of the system. For the assumed parameters, we can compute $\gamma = 11.204$, $\rho_k = 0.474$, $\rho_c = 0.836$, $K_e/m = 11.609\ 1/s^2$, and $S_e/m = 13.041\ 1/s^2$. The stationary root mean square displacement of the equivalent linear hysteretic system, σ_{y_1}, is computed from Eqs. (92) and (93) as

$$\sigma_{y_1} = 2.70\ cm \tag{96}$$

Comparing Eqs. (95) and (96) we see that the accuracy of the linearization method is excellent. Figure 18 compares the results obtained in the Monte Carlo simulation with the stationary value estimated by means of the hysteretic linearization procedure.

Although a frequency-domain approach has been suggested above, the following time-domain representation can be used for linear hysteretic damping[16]

$$f(t) = K_e \, \Delta(t) + S_e \, \hat{\Delta}(t) \tag{97}$$

where $\hat{\Delta}(t)$ is the Hilbert transform of $\Delta(t)$ defined as

$$\hat{\Delta}(t) = -\frac{1}{\pi} \int_{-\infty}^{\infty} \frac{\Delta(\tau)}{t - \tau} \, d\tau \tag{98}$$

In Eq. (98) the Cauchy principal value of the integral (see Appendix C) is implied. It is worth mentioning that, although the Hilbert transform allows a time-domain representation for structures with linear hysteretic damping, the solution in the frequency domain is more convenient than solutions in the time domain because of the non-causality property of the Hilbert transform. However, iterative techniques in the time domain or state-space methods with analytic signals can be used to compute the response of linear structures with linear hysteretic elements[13,16].

5. Experimental research

This section is devoted to the verification of the mathematical models proposed in previous sections for the EDR dampers. The response of a base-isolated test structure with EDR dampers and a mass damper device subjected to simulated ground motions on a shake table is estimated integrating the equations of motion of a mathematical model of the test structure and compared with the recorded responses.

5.1. Earthquake simulator tests on an isolated model

Testing of a reduced-scale structure with EDR dampers was performed recently[22]. Two applications of the EDR were considered during this experimental program: (i) the use of the EDR as a base-isolation system and (ii) its use as an energy absorbing strut for building structures. A steel moment-resistant frame was used as a test structure for both applications. A detailed description of the objectives and results of this experimental program is given elsewhere[22]. Here, a mathematical model for the test structure is obtained and the response of the test structure containing EDR dampers is computed and correlated with the corresponding experimental results.

A schematic of the configuration used to test the EDR damper as a component of an isolation system is shown in Fig. 19. This figure shows the three-story, moment-resistant frame structure used in the experimental program. The test frame was 6 ft tall and 3 ft by 4 ft in plan. Approximately 1000 lb of lead were added to each floor and to the base of the structure. The model was mounted on rollers on a shaking table capable of providing horizontal motion in one direction. The relative motion of the base of the structure with respect to the shake table was constrained by the rollers in all but a single horizontal direction parallel to the motion of the shake table. The base of the isolated structure was connected to the shake table using an EDR damper which functioned as an isolation system in the horizontal direction. The EDR was configured so that deformations of up to ± 5 *in* could be accommodated in the isolation system.

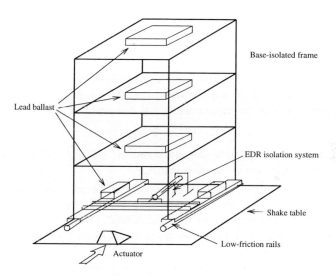

Figure 19. Schematic view of the experimental setup.

The instrumentation was designed to capture the motion of the floors of the structure and the force in the EDR damper. Accelerometers were located on each floor of the structure, on the base of the structure and on the shake table. Linear potentiometers were used to measure the displacements of the floors of the structure, and direct current differential transducers were used to measure the deformation of the EDR damper and the displacement of the table.

A four-degree-of-freedom lumped-parameter model was identified for the base-isolated model. The selected degrees of freedom were $q_1(t)$ = displacement of the base of the structure, $q_2(t)$ = displacement of the first floor, $q_3(t)$ = displacement of the second floor, and $q_4(t)$ = displacement of the fourth floor.

From weight tests performed on the structure, the weights of each floor and base were estimated as $W_{base} = 1270$ *lb*, $W_1 = 1324$ *lb*, $W_2 = 1342$ *lb*, $W_3 = 1301$ *lb*. Accordingly, the mass matrix was taken as

$$\mathbf{M} = \begin{bmatrix} 3.288 & 0 & 0 & 0 \\ 0 & 3.428 & 0 & 0 \\ 0 & 0 & 3.474 & 0 \\ 0 & 0 & 0 & 3.368 \end{bmatrix} \; lb \; s^2/in \tag{99}$$

Using recorded data from free vibration and impact tests of the model isolated with linear springs, the stiffness matrix of the superstructure associated with the horizontal motions of the floors and the base was estimated. The stiffness matrix of the superstructure (excluding the isolators) was estimated as

$$\mathbf{K} = \begin{bmatrix} 13441 & -20733 & 8957 & 1511 \\ -20733 & 42082 & -29805 & 8456 \\ 8957 & -29805 & 37267 & -16419 \\ -1511 & 8456 & -16419 & 9474 \end{bmatrix} \ lb/in \quad (100)$$

Modal damping ratios of $\xi_i = 0.005$ were estimated for the three vibration modes that involved deformation of the superstructure (i.e., for modes $i = 2, 3, 4$). Accordingly, the damping matrix of the superstructure was computed as

$$\mathbf{C} = \mathbf{\Phi}^{-T} \ \mathbf{\Lambda} \ \mathbf{\Phi}^{-1} \quad (101)$$

where $\mathbf{\Lambda}$ is a diagonal matrix with diagonal terms $\Lambda_{ii} = 2\xi_i \omega_i$ and $\mathbf{\Phi}$ is the modal matrix normalized with respect to the mass matrix.

A frictional force of approximately 20 *lb* existed in the rail supporting the sliders of the isolated structure. Although small (less than 0.5% of the model weight), this frictional force affected the response of the base-isolated model. For that reason, this dry friction was accounted for in the mathematical model of the structure.

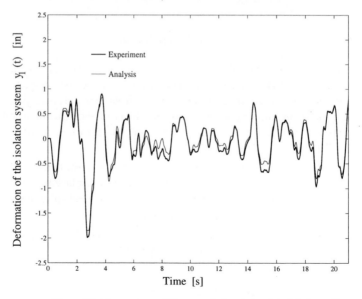

Figure 20. Comparison of the recorded and simulated deformation.

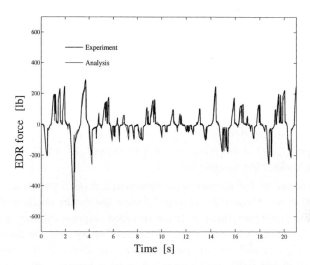

Figure 21. Comparison of recorded and simulated EDR force.

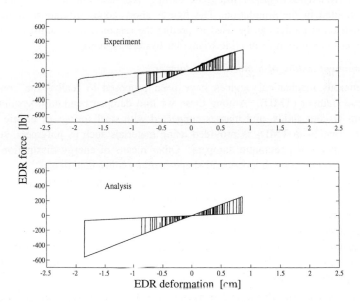

Figure 22. Comparison of recorded and simulated hysteresis loops.

In the experiments the model was isolated using a single EDR damper and subjected to support motion simulating a recorded earthquake (a signal recorded at the coastal town of Zacatula during the 1985 Michoacan earthquake). The recorded response of the model isolated with an EDR device was later compared with the prediction obtained using the mathematical model of the superstructure described above and the linear-friction element described in section 2. The following model was used for the structure isolated with the EDR damper:

$$\mathbf{M}\,\ddot{\mathbf{y}}(t) + \mathbf{C}\,\dot{\mathbf{y}}(t) + \mathbf{K}\,\mathbf{y}(t) + \mathbf{L}^T(f_{EDR}(t) + f_{df}(t)) = -\mathbf{M}\,\mathbf{l}\,w(t) \qquad (102)$$

where $\mathbf{M}$ is given in Eq. (99), $\mathbf{K}$ in Eq. (101), $\mathbf{C}$ in Eq. (102), $f_{EDR}(t)$ is the force in the EDR damper, $f_{df}(t)$ is the frictional force in the rail, $\mathbf{L}^T = [1\ 0\ 0\ 0]$, $\mathbf{l} = [1\ 1\ 1\ 1]^T$, and $w(t)$ is the shake table acceleration.

The parameters of the damper were determined as $K_1 = 301\ lb/in$, $K_2 = 37\ lb/in$, and $K_3 = 10000\ lb/in$. Figures 20 through 22 show the results obtained in the numerical simulation and compare them with the recorded responses during the test. Figure 20 shows the deformation history, $y_1(t)$, of the isolation system; the recorded response is shown in thick line and the computed response is shown in thin line. Figure 21 shows the force in the EDR device recorded using a load cell during the test (thick line) and the force history obtained in the numerical simulation (thin line). Figure 22 shows the hysteresis loops of the EDR damper recorded during the experiment and those computed in the simulation. The figures show good correlation, indicating that the linear-friction model can be used to predict the response of structures incorporating EDR devices configured to provide triangular hysteresis loops.

5.2. Component testing of a tuned mass damper

Numerous mechanical devices have been proposed to realize the concept of a tuned mass damper (TMD). Among these we find damped pendulum systems, vessels filled with viscous fluids, and masses connected with steel springs or elastic rods. The energy dissipation in TMDs is provided using materials such as polymers, steel ropes, friction dampers, or pneumatic dampers. Other means of energy dissipation proposed include electro-magnetic dampers and shape-memory alloy materials.

For efficiency in reducing vibrations in linear structures, a TMD must have period of vibration and decay ratio independent of the amplitude of oscillation, should accommodate for the need of fine tuning after installation, and should have stable mechanical properties during operation. Although linear viscoelastic materials are commonly used in the resistance scheme of the TMD, analytical research has suggested that nonlinear mechanisms exhibiting triangular hysteresis loops under cyclic loading are suitable for mass damping applications[11,15].

To investigate the dynamical behavior of a TMD with EDR dampers an experimental program was conducted. A mass damping device was constructed by mounting a mass on frictionless bearings and attaching it to a shaking table with helical springs

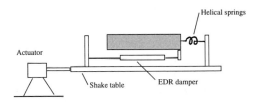

Figure 23. Schematic of the TMD with an EDR damper.

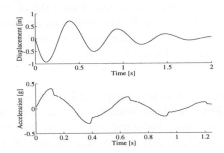

Figure 24. Deformation and acceleration of the TMD recorded
in a free vibration test.

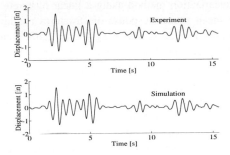

Figure 25. Comparison of experimental and simulated TMD deformations.

and an EDR damper for testing (Fig. 23). The weight of the mass was approximately 320 *lb* and the springs and EDR were designed such that upon loading the stiffness of the connection was $K_1/m = 12.7^2 \ 1/s^2$ and upon unloading the stiffness of the connection was $K_2/m = 10.17^2 \ 1/s^2$, where m is the mass of the TMD.

Free vibration and forced vibration tests were conducted on the model. Figure 24 shows the deformation and acceleration histories recorded in a free vibration test. As it can be observed in the figure, the period of vibration and the decay ratio of this nonlinear oscillator are independent of the amplitude of vibration, confirming the analytical predictions made in Eqs. (33) and (34). The recorded TMD deformation history during a test simulating the El Centro earthquake (scaled to 1.3 inches maximum displacement) is shown in Fig. 25. In the same figure, the simulated result is presented which shows excellent agreement with the experimental result.

The good correlation obtained between the numerical simulations and the experimental results provides great confidence in the use of the EDR models developed. The main implication of these results is that further studies on the behavior of structures with EDR dampers can be developed using computer models. Such studies can include linear and nonlinear models for the main structural system.

6. Concluding remarks

The dynamical response of structures containing EDR dampers configured to provide triangular hysteresis loops has been investigated. The free vibration response of SDOF and MDOF structures with linear-friction dampers has been studied. These structures show modes of vibration, period of oscillation and decay ratio independent of the amplitude of vibration, quite uncommon features for nonlinear structures. The harmonic linearization technique has shown excellent accuracy in the estimation of the response of structures with EDR dampers subjected to harmonic and broad-band excitation. The MSE method in combination with the harmonic linearization technique has been used to estimate the dynamical response of MDOF structures containing linear-friction elements and a linearization method using a linear hysteretic element has been proposed to estimate the mean square response of structures containing linear-friction elements subjected to random excitation. By using Monte Carlo simulation techniques, the accuracy of this method has been demonstrated to be excellent.

The EDR damper when used as an energy dissipation mechanism in a vibrating system or as a device for the enhancement of the seismic performance of a structural system at first sight may appear to be less effective than a conventional Coulomb friction damper in the sense that it does not dissipate as much energy in a cycle for the same force level. However, it has advantages over the Coulomb friction devices including the fact that it leads, as shown here, to highly accurate equivalent linearization techniques. It also has the characteristic that it dissipates energy over the entire range of excitation intensity. In a conventional element where there is a threshold force that must be overcome before the device becomes effective, the level of input must be predicted because the dynamics of the structure with supplemental Coulomb friction dampers depends highly on the response amplitude; if the threshold is set too high in order to have adequate damping at a high level of input, the device will be ineffective at lower levels of input. In contrast, a linear-friction device such as the EDR damper will be active at all levels of input.

Because the dissipation of energy-per-cycle of this nonlinear homogeneous element is quadratic in the deformation amplitude, excellent accuracy is obtained with the linearization methods. In fact, the accurate predictability of the response of structures containing EDR elements with triangular hysteresis loops is a very convenient characteristic of this type of energy dissipator. The linearization techniques applied in this work are useful tools for the preliminary design of structures with EDR dampers.

7. Acknowledgments

Support for different phases of this work has been provided by the National Center for Earthquake Engineering Research under Grants No. NCEER 93-5201A and NCEER 94-5201A, and by the National Science Foundation under Grant No. BCS 9302101. Fluor Daniel Inc. provided the dampers used in the experiments reported. This support is greatly appreciated. Doug K. Nims provided part of the experimental data used herein for model validation.

8. References

1. I. A. Aiken and J. M. Kelly, *Earthquake simulator testing and analytical studies of two energy-absorbing systems for multistory buildings,* Report No. UCB/EERC-90/03, of California at Berkeley, (1990).

2. R. E. Bishop, 1955, *J.R. Aero. Soc.,* **59** (1955), p. 738-742.

3. R. N. Bracewell, *The Fourier transform and its applications,* (McGraw-Hill, 1986).

4. T. K. Caughey and A. Vijayaraghavan, *Int. J. Non-Linear Mechanics,* **5** (1970), p. 533-555.

5. T. K. Caughey and A. Vijayaraghavan, *Int. J. Non-Linear Mechanics,* **12** (1977), p. 339-353.

6. D. C. Champeney, D.C., *A handbook of Fourier theorems,* (Cambridge University Press, Great Britain, 1987).

7. S. H. Crandall, *Air, Space and Instruments, (L. Sidney, Ed.) (McGraw-Hill Book Company Inc., 1963).*

8. T. F. Fitzgerald, T. Ahagnos, M. Goodson, and T. Zsutty, *Earthquake Spectra,* **5** (1989), p. 383-391.

9. T. Fujita, (editor), *Seismic isolation and response control for nuclear and non-nuclear structures,* 11th Int. Conf. on Str. Mechanics in Reactor Technology, SMIRT, Tokyo, Japan, 1992.

10. C. E. Grigorian, T. Shuoh-Yang, and E. P. Popov, *Slotted bolted connection energy dissipators,* Report No. UCB/EERC-92/10, of California at Berkeley, (1992).

11. J. A. Inaudi and J. M. Kelly, *A friction mass damper for vibration control,* Report No. UBC/EERC-92/15, Earthquake Engineering Research Center, University of California at Berkeley, 1992.

12. J. A. Inaudi, G. Leitmann, G., and J. M. Kelly, *J. Eng. Mechanics,* **120** (1994), p. 1543-1562.

13. J. A. Inaudi and N. Makris, *Time-domain analysis of linear hysteretic damping,* (Submitted for publication, 1995).

14. J.A. Inaudi, D.K. Nims, and J.M. Kelly, *On the analysis of structures with energy dissipating restraints,* EERC Report No. 93/13, Earthquake Engineering Research Center, University of California at Berkeley, 1993.

15. J. A. Inaudi and J. M. Kelly, *Proc. Ninth VPI&SU on Dynamics and Control of Large Structures* (Ed. L. Meirovitch), pp. 531-542, Blacksburg, Virginia, 1993.

16. J. A. Inaudi and J. M. Kelly, Linear hysteretic damping and Hilbert transform, *Journal of Engineering Mechanics,* **121** (1995).

17. J.A. Inaudi, D.K. Nims, and J.M. Kelly, *Proc. 1994 North American Conference on Smart Structures and Materials, SPIE Proceedings Vol. 1193* (Ed. C. Johnson), p. 213-224, (Int. Society for Optical Engineering, 1994).

18. W. D. Iwan, *Int. J. Non-Linear Mechanics,* **8** (1973) p. 279-287.

19. C. D. Johnson and D. A. Kienholz, *AIAA Journal,* **20** (1982), p. 1284-1290.

20. B. G. Korenev and L. M. Reznikov, *Dynamic Vibration Absorbers,* (John Wiley & Sons, Chichester, England, 1993).

21. P. Lancaster, *J.R. Aero. Soc.,* **64** (1960), p. 229.

22. D. K. Nims and J. M. Kelly, *Introduction to the small scale seismic testing of a self-centering friction energy dissipator for structures,* Report to Sponsor, Department of Civil Engineering, University of California at Berkeley, 1990.

23. D. K. Nims, P. J. Richter, and R. E. Bachman, *Earthquake Spectra,* **9** (1993), p. 467-489.

24. A. S. Pall and C. March, 1982, *J. Structural Engineering, ASCE,* **108** (1982), p. 1213-1323.

25. A. Pall, S. Venzina, P. Proulx, and R. Pall, *Earthquake Spectra,* **9** (1993), p. 547-557.

26. T. J. Reid, 1956, *J.R. Aero. Soc.,* **60** (1956) p. 283.

27. P. J. Richter, D. K. Nims, J. M. Kelly, and R. M. Kallenbach, 1990, *Proc. of the 59 Annual Convention,* Structural Engineers Association of California, **1** (1990), p. 378-401.

28. J. B. Roberts, J.B. and P. D. Spanos, *Random Vibration and Statistical Linearization,* (John Wiley & Sons, New York, 1990).

29. R. Shepperd and L. A. Erasmus, *Proc. Ninth World Conference on Earthquake Engineering,* **V** (1988), p. V-767-772.

30. Soong, T.T., and Constantinou M.C., (Eds.), *Passive and active structural vibration control in civil engineering,* (Springer Verlag, New York, 1994).

31. E. E. Ungar and E. M. Kerwin, *J. of Acoustical Society of America,* **34** (1962), p. 954-957.

Appendix A: Free vibration of a SDOF system with EDR damper

Figure 10 describes the free vibration of the SDOF oscillator in the phase plane. There are three switching lines that separate three regions corresponding to loading, unloading and transition between loading and unloading. Two of these switching lines coincide with the horizontal and vertical axes ($y(\theta) = 0$ and $y'(\theta) = 0$). In this Appendix, the solution of Eq. (27) for the initial conditions $y(0) = Y_n$, $y'(0) = 0$, and $y_l(0) = z(0) = Y_n$ is computed to obtain the slope of the third switching line.

The phase of the solution corresponding to the transition between loading and unloading is governed by the following differential equation

$$y''(\theta) + (1+\gamma \zeta) y(\theta) = Y_n \zeta (\gamma-1) \tag{103}$$

where

$$\zeta = \frac{K_h}{\omega^2 m} \tag{104}$$

and γ is given in Eq. (23). The solution can be written as

$$y(\theta) = Y_n \frac{\zeta (\gamma-1)}{1 + \gamma\zeta} + Y_n (1-\frac{\zeta (\gamma-1)}{1 + \gamma\zeta}) \cos(\sqrt{1 + \gamma\zeta}\, \theta) \tag{105}$$

$$y'(\theta) = -Y_n \frac{1 + \zeta}{\sqrt{1 + \gamma\zeta}} \sin(\sqrt{1 + \gamma\zeta}\, \theta) \tag{106}$$

At $\theta = \theta_1$ the system reaches the unloading phase; i.e.,

$$y(\theta_1) = Y_n \frac{\gamma-1}{\gamma+1} \tag{107}$$

From Eqs. (105) and (107) we obtain

$$\theta_1 = \frac{1}{\sqrt{1 + \gamma\zeta}} \cos^{-1}[(\frac{\gamma-1}{\gamma+1} - \frac{\zeta(\gamma-1)}{1 + \gamma\zeta}) \frac{1 + \gamma\zeta}{1 + \zeta}] \tag{108}$$

The sought switching line has a slope $\dfrac{y'(\theta_1)}{y(\theta_1)}$ in the phase plane given by

$$\frac{y'(\theta_1)}{y(\theta_1)} = -\frac{\gamma + 1}{\gamma - 1} \frac{(1+\zeta) \sin(\sqrt{1 + \gamma\zeta}\theta_1)}{\sqrt{1 + \gamma\zeta}} \tag{109}$$

An alternative way to compute the slope of the switching line approach follows. Using the fact that $y(\theta)'' = \dfrac{dy'(\theta)}{dy(\theta)} y'(\theta)$, and using Eq. (103), we can write

$$\frac{dy'(\theta)}{dy(\theta)} y'(\theta) + y(\theta) (1 + \gamma\zeta) = Y_n \zeta (\gamma-1) \tag{110}$$

whose solution can be written as

$$y'(\theta)^2 + (1 + \gamma\zeta) y(\theta)^2 = 2Y_n \zeta (\gamma-1) + [(1 + \gamma\zeta) - 2\zeta (\gamma-1)] Y_n^2 \tag{111}$$

On the switching line $y(\theta_1) = \frac{\gamma-1}{\gamma+1} Y_n$, and, therefore, using this expression and Eq. (111), we can write

$$\frac{y'(\theta_1)}{y(\theta_1)} = -\frac{\gamma+1}{\gamma-1} \left[2\zeta \frac{(\gamma-1)^2}{\gamma+1} + 1 + \gamma\zeta - 2\zeta(\gamma-1) - (1+\gamma\zeta)(\frac{\gamma-1}{\gamma+1})^2 \right]^{\frac{1}{2}} \quad (112)$$

Both Eqs. (109) and (112) render the same numerical values for the slope of the switching line.

Appendix B: Harmonic linearization of a linear-friction element

Consider Eq. (58) with $\Delta(t) = \sin \bar{\omega} t$. Let $\theta = \bar{\omega} t$, then the harmonic linearization method reduces to

$$minimize \quad J(c_e, k_e) = 2 \int_0^\pi [\ k_e\ \Delta(\theta) + c_e\ \dot{\Delta}(\theta)) - g(\Delta(\theta), l(\theta), z(\theta))]^2\ d\theta \qquad (113)$$

where, for $\Delta(\theta) = \sin(\theta)$, $g(\Delta(\theta), l(\theta), z(\theta))$ is given by

$$g(\theta) = \begin{cases} K_h \sin\theta\ , & 0 < \theta < \dfrac{\pi}{2} \\[2mm] K_h(1-\gamma+\gamma\sin\theta)\ , & \dfrac{\pi}{2} < \theta < \dfrac{\pi}{2} + \cos^{-1}(\dfrac{\gamma-1}{\gamma+1}) \\[2mm] -K_h \sin\theta\ , & \dfrac{\pi}{2} + \cos^{-1}(\dfrac{\gamma-1}{\gamma+1}) < \theta < \pi \end{cases} \qquad (114)$$

The equivalent constants can be computed as

$$k_e = \frac{\displaystyle\int_0^\pi g(\theta)\ \sin\theta\ d\theta}{\displaystyle\int_0^\pi \sin^2\theta\ d\theta} \qquad (115)$$

$$c_e = \frac{\displaystyle\int_0^\pi g(\theta)\ \cos\theta\ d\theta}{\bar{\omega}\displaystyle\int_0^\pi \cos^2\theta\ d\theta} \qquad (116)$$

Solving the integrals in Eqs. (115) and (116) in the intervals $0 < \theta < \dfrac{\pi}{2}$, $\dfrac{\pi}{2} < \theta < \dfrac{\pi}{2} + \cos^{-1}(\dfrac{\gamma-1}{\gamma+1})$, and $\dfrac{\pi}{2} + \cos^{-1}(\dfrac{\gamma-1}{\gamma+1}) < \theta < \pi$, the following expressions are obtained

$$k_e = \frac{K_h}{\pi}\ [\ \cos^{-1}(\frac{\gamma-1}{\gamma+1})\ (1+\gamma) - 2\sqrt{\gamma}\ \frac{\gamma-1}{\gamma+1}] \qquad (117)$$

$$c_e = \frac{K_h}{\bar{\omega}\,\pi}\ [\ 1 + 4\frac{\gamma-1}{\gamma+1} - \frac{4\gamma^2}{(\gamma+1)^2} + (\frac{\gamma-1}{\gamma+1})^2\ ] \qquad (118)$$

Appendix C: Linear hysteretic damping and the Hilbert transform

The mechanical-element model discussed herein shows frequency-independent storage and loss moduli; this implies that the Fourier transform of the element force $F(j\varpi)$ and Fourier transform of the element deformation $\Delta(j\varpi)$, satisfy:

$$F(j\varpi) = k(1 + j\ \eta\ sgn(\varpi))\ \Delta(j\varpi) \tag{119}$$

where k is a parameter with stiffness units, $j = \sqrt{-1}$, η is the frequency-independent loss factor (ratio of the loss and storage moduli of the element), and $sgn(.)$ is the signum function, $sgn(x) = 1$ if $x > 0$, $sgn(x) = -1$ if $x < 0$, and $sgn(x) = 0$ if $x = 0$.

The Hilbert transform

In this section, it is shown that the Hilbert transform gives a correct time-domain representation for the linear hysteretic damping model defined in Eq. (119). First, let us recall the definition of this transform and briefly look at some of its properties.

The functions $g(t)$ and $\hat{g}(t)$ are called a Hilbert transform pair if, for almost all t,

$$\hat{g}(t) = \lim_{a \to \infty} P \int_{-a}^{a} \frac{-g(\tau)}{\pi(t - \tau)}\ d\tau \tag{120}$$

$$g(t) = \lim_{a \to \infty} P \int_{-a}^{a} \frac{\hat{g}(\tau)}{\pi(t - \tau)}\ d\tau \tag{121}$$

where

$$P \int_{-\infty}^{\infty} \frac{f(\tau)}{t - \tau}\ d\tau = \lim_{\varepsilon \to 0^+} \left[\int_{t+\varepsilon}^{\infty} \frac{g(\tau)}{t - \tau}\ d\tau + \int_{-\infty}^{t-\varepsilon} \frac{g(\tau)}{t - \tau}\ d\tau \right] \tag{122}$$

is called the Cauchy principal value around $\tau = t$ of the integral[6]. We say that $\hat{g}(t)$ is the Hilbert transform of $g(t)$, and we write $\hat{g}(t) = H[g(t)]$. For these integrals to converge, $f(t)$ must be p-th power Lebesgue integrable, for some p in the interval $(1, \infty)$. The key feature of the Cauchy principal value is that, by taking the limits for $\varepsilon \to 0^+$ in both integrals, the infinities generated to the right and to the left of $\tau = t$ can cancel each other in continuous signals.

As shown in Eq. (120), the Hilbert transform of a signal $g(t)$ is a linear operator defined by the convolution of $g(t)$ and $-1/(\pi t)$. Equation (120) also shows that this operator is non-causal, since in order to compute $\hat{g}(t)$, the function $g(\tau)$ is required for $-\infty < \tau < \infty$.

From Eq. (120) we note that the Hilbert transform of a constant signal is zero. The Hilbert transform is invariant under time scaling[3]; this means that the Hilbert transform of $g(\alpha t)$ is $\hat{g}(\alpha t)$ (similarity property) where $\hat{g}(t)$ is the Hilbert transform of $g(t)$. Consider the signal $g(t) = \sin \varpi t$ with $\varpi \neq 0$; its Hilbert transform is

$$H[\sin \bar{\omega} t] = \int_{-\infty}^{\infty} \frac{-\sin \bar{\omega} \tau}{\pi (t - \tau)} \, d\tau \tag{123}$$

Letting $x = \tau - t$ and expanding the sine of a sum, we obtain:

$$H[\sin \bar{\omega} t] = \cos \bar{\omega} t \ \frac{1}{\pi} \int_{-\infty}^{\infty} \frac{\sin \bar{\omega} x}{x} \, dx = \cos \bar{\omega} t \ sgn(\bar{\omega}) \tag{124}$$

Similar operations yield

$$H[\cos \bar{\omega} t] = -\sin |\bar{\omega}| t \tag{125}$$

As shown in Eqs. (124) and (125), the Hilbert transform does not change the amplitude of a sine or cosine signal and only changes its phase by $\pm \pi/2$ *rad*.

To further investigate the Hilbert transform, a frequency-domain analysis is useful. Applying Fourier transform to the convolution defined in Eq. (120) we obtain

$$\hat{G}(j\bar{\omega}) = FT[\hat{g}(t)] = FT[-\frac{1}{\pi t}] \ G(j\bar{\omega}) = j \ sgn(\bar{\omega}) \ G(j\bar{\omega}) \tag{126}$$

where $FT[.]$ represent Fourier transformation. Therefore, a time-domain representation for linear hysteretic damping can be obtained using the Hilbert transform

$$f(t) = k \ \Delta(t) + k \ \eta \ \hat{\Delta}(t)$$

Dynamics with Friction: Modeling, Analysis and Experiment, pp. 137–168
edited by A. Guran, F. Pfeiffer and K. Popp
Series on Stability, Vibration and Control of Systems Series B: Vol. 7
© World Scientific Publishing Company

FRICTION AND IMPACT DAMPING
IN A TRUSS USING PINNED JOINTS

STEVEN FOLKMAN, BROOK FERNEY, JEFFREY BINGHAM AND JOSEPH DUTSON
Mechanical and Aerospace Engineering, Utah State University
Logan, Utah 84322-4130, USA

ABSTRACT

Researchers at Utah State University have been investigating mechanisms involved
with structural damping in a truss whichis attributable to pinned joints. The joints
investigated are typical clevis/tang joint with clearance fit pins. A review of work
discussing damping mechanicsms is given. Expressions for damping due to friction
caused by either extensional or rotational motions are derived. Mechanisms for
damping caused by impacting are discussed. Measured results for a truss with
eight pinned joints are given and effects of gravity on damping are demonstrated.
The role of friction and impacting on damping in the truss is summarized. A simple
nonlinear finite element model of a joint is discussed.

1. Introduction

The design of a deployable space structure typically incorporates multiple
revolute joints into a deployable truss. Joint designs are largely influenced by the
methods used to construct the truss in space. If the truss is constructed by deployment or
unfolding of truss components, the joints may utilize pinned joints. Figure 1 illustrates a
deployable joint design proposed for large reflector antenna structures.[1] A pinned-joint
design would allow rotational slippage about the pin axis. This rotational slippage could
be exploited to maximize joint damping by controlling the loading of the joint pins and
running the joints dry (unlubricated). Drawbacks of this approach include permanent
offset of the truss due to residual Coulomb friction and fretting corrosion of the joints.
Utilizing a viscoelastic material in the joints could provide a source of viscous damping.
A major current drawback of joint design such as that illustrated in Figure 1 is the
inability to accurately predict the dynamic behavior of the truss. Because the structure
may exhibit nonlinear behavior, such designs are often avoided due to the difficulty in
modeling the behavior.

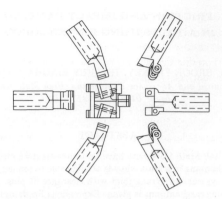

Figure 1   Illustration of a joint for a deployable structure.

If these joints are designed such that large preloads can be applied across the mating surfaces or preloaded bearings are installed, the damping produced by that joint approaches that of a welded or tightly clamped joint. The contribution of the joints to structural damping would be small and linear finite element models would normally be able to accurately predict the behavior of these structures. However, if the joint design allows a small amount of "slop" or deadband, the dynamic behavior of this structure can be dramatically altered. The deadband and friction characteristics in this type of joint can introduce nonlinearities into the behavior. The structure would also exhibit deadband, hysteresis, and have significantly higher damping rates than identical structures with "welded" joints. Unfortunately, dynamic behavior due to the joints is very dependent on many variables such as joint design and condition of joint interfaces. Predicting the dynamic behavior is at best difficult. Measured data from design prototypes is often required. The problem with large scale testing is that testing usually occurs at a time when most of the structural design has been fixed in detailed design drawings or testing in a 1-g environment may be difficult for structures not designed to operate under that load.

2. Passive Damping in Space Structures

Ashley[2] noted that the primary sources for passive damping for space structures could be fit into three categories:
1) material damping;
2) damping at joints and interconnections and;

3) artificially introduced damping (dashpots).
For small amplitudes, material damping can often be modeled as being independent of stress level. Such is not the case for joint damping. Simple models (i.e., Coulomb friction with macro slip) predict the rate of energy dissipation to be dependent on the normal loads across the interface of a joint and the magnitude of the relative motion in the joint interface. This infers that gravity should influence damping rates. If joints allow some slippage, a 1-g load could prevent or reduce the amount of slippage which would occur and thus reduce damping. However, if a fixed amount of slippage occurs in a joint and joint loads are increased, damping would also increase. Thus, preloads due to gravity could either increase or decrease damping. Ground tests of lightweight structures may be significantly influenced by gravity induced loads.

The amount of damping which joints could provide for space structures is uncertain. Ashley[2] postulates that material damping will dominate over joint damping for very large space structures. However, his conclusion is based upon the assumption that joint damping is proportional to a characteristic length of the structure squared. Generally, joint damping via Coulomb friction is dependent on the number of joints, which is proportional to structure volume as is material damping. If the lengths of individual truss members are held constant, the relative contributions of material and joint damping should not change with structure size. Plunkett[3] gave a summary of friction damping knowledge and reported that "several studies have shown that joint and connection damping is the most important mechanism for energy dissipation in most real structures."

Folkman[4] conducted ground tests of a miniature truss constructed using pinned joints and demonstrated that damping rates could change by a factor of 4 due to gravity loads. It is suggested that, depending on the type of joints selected, the magnitude and character of structural damping can be significantly influenced by joints. If joints are designed such that large preloads can be applied across the mating surfaces, the damping produced by those joints approaches that of a welded joint. However, joints which have a small amount of deadband can influence structural dynamics significantly.

3. Reported Joint Damping Mechanisms

Models of joint or interface damping are derived from two mechanisms: friction and impacting. Friction is attributed to either rotary or extensional motions; impacting implies that two surfaces are separated by some finite gap and come into contact during each cycle of an oscillation.

The method of energy dissipation for impacting is rather complicated. Crawley et al.[5] suggested that one measure of energy dissipation due to impacting is the coefficient of restitution. The coefficient of restitution is not only a material property but is also a function of the shape of the contacting surfaces.

Models of friction damping generally fall into two categories: microslip and macroslip. Macroslip models assume no damping occurs until there is relative motion between two interfaces. Relative motion occurs when the forces parallel to the interface exceed the Coulomb frictional force, which is proportional to the force which is normal to the interface. This classic friction-damping model was analyzed by Den Hartog.[6] His analysis indicates that for small loads, the energy dissipated increases linearly with displacement. Beards and Williams[7] discussed how to optimize joint damping by maintaining an optimum joint load during rotational macroslip. They reported that significant damping rates can be obtained when joints are allowed to undergo rotational slippage. However, some static stiffness is sacrificed when rotational slip is allowed.

Microslip models predict friction damping due to localized, microscopic slippage. Because of surface imperfections, interface contact pressure is not uniformly distributed. This allows localized slippage, while the overall joint remains "locked." For example, when material damping measurements are made using cantilever beam specimens, a prime concern is how the specimen is clamped to the "wall" such that microslip contributions are minimized. Damping due to microslip would typically be less than from macroslip. Plunkett[3] reported we are far from being able to predict damping due to microslip.

No general model is available describing losses in joints. However, analytical and computer models which attempt to include friction and impacting of a structure containing a single joint are available.[8-11] Ferri[11] showed computer simulations of the behavior of a single joint indicating that nonlinear sources of damping, such as friction and impacting, appear to be predominately viscous in nature. However, the models in the above references are not directly applicable to truss structures with multiple joints, although they could provide damping estimates of individual components. No references were found comparing model results with measured data.

3.1 Commonly Used Terms

Commonly used terms describing damping in a structural component include the loss factor η which is defined as:

$$\eta = \frac{\Delta U}{2\pi U} \tag{1}$$

where ΔU is the energy dissipated per cycle and U is the energy stored per cycle. The viscous damping ratio ζ is defined by:

$$\zeta = \frac{c}{c_{cr}} \qquad (2)$$

where c is the viscous damping coefficient and c_{cr} is the critical viscous damping coefficient. A common method used to measure damping of a single degree of freedom system is to record the rate of decay of a single vibration mode. The logarithmic decrement δ is then computed from:

$$\delta = \frac{1}{n} \ln\left(\frac{A_0}{A_n}\right) \qquad (3)$$

where n is the number of cycles of the decay and A_0 is amplitude of displacement at some initial reference while A_n is the amplitude after n cycles. Another term describing damping is the amplitude ratio at resonance or Q which can be defined easily in terms of ζ as:

$$Q = \frac{1}{2\zeta} \qquad (4)$$

Relationships between δ and ζ are given by:

$$\delta = \frac{2\pi\zeta}{\sqrt{1-\zeta^2}} \quad \text{or} \quad \zeta = \frac{\delta}{\sqrt{4\pi^2 + \delta^2}} \qquad (5)$$

For light damping (i.e., $\delta < 0.01$), η, ζ, δ, and Q can be simplified to the following relationships:

$$\eta \approx 2\zeta \approx \frac{\delta}{\pi} \approx \frac{1}{Q} \qquad (6)$$

Although the above terms are all commonly used, the relationships among the terms are strictly correct only for viscous damping (when damping forces are proportional to the structure velocity). Material damping at low stress levels for most structural materials can often be modeled as viscous damping. Damping due to frictional effects is usually assumed to be independent of velocity and is proportional to the normal forces between the contacting surfaces. Damping due to joints and connections is usually best described in terms of η, the loss factor.

3.2 Damping in Built-up Structures

For a structure built from several components, a loss factor for the entire structure, η_t, can be obtained by summing the energy dissipated in each element and dividing by the total strain energy in the structure, or:

$$\eta_t = \frac{\sum\limits_{i=1}^{n} \Delta U_i}{2\pi \sum\limits_{i=1}^{n} U_i} \tag{7}$$

where ΔU_i is the energy dissipated in element i and U_i is the energy stored in element i. Let U_t be the total strain energy in a structure and f_i be the fraction of the total strain energy in component i or:

$$U_t = \sum_{i=1}^{n} U_i \tag{8}$$

$$f_i = \frac{U_i}{U_t} \tag{9}$$

then η_t can be expressed as:

$$\eta_t = \sum_{i=1}^{n} \frac{\Delta U_i}{2\pi U_i} \frac{U_i}{U_t} = \sum_{i=1}^{n} \eta_i f_i \tag{10}$$

Thus, the loss factor for a structure could in principal be computed from loss factors of the individual components multiplied by the fraction of the total strain energy stored in that component. Furthermore, if the loss factors were assumed to be constant for all components, then the components which store a small fraction of the total strain energy (such as a joint) would contribute less to the total structural damping. This does indicate that a mechanism or treatment used to enhance damping should be located in high strain energy regions of the structure.

4. Damping Models

Simple models can be derived which can provide qualitative and some quantitative information about damping due to friction and impact. Damping due to friction can be divided into two categories: friction due to extensional and to rotary displacements. It is assumed here that friction forces are proportional to the normal force at an interface (i.e. Coulomb Friction Model). More elaborate friction models are available[12] which include the effect of velocity on friction force.

4.1 Extensional Friction Damping Model

The following is a derivation of the loss factor due to extensional displacements in a strut of length L containing a pinned joint at each end. Consider the joint illustrated in Figure 2. The pin in the end of the strut is greatly enlarged to allow the contact points to be more easily seen. The pin is illustrated as having an octahedral shape; however, the angle β is variable. The pin is assumed to be rigidly attached to the strut and contacts the joint at the blocks oriented at angle β. The stiffness of each joint is modeled using four springs with stiffness k_j. The strut stiffness is k_s. The strut carries an initial load F_i and a load F which varies as a sinusoid with time. A perfect fit of the pin to the joint (zero deadband) is assumed. That is, when the axial force in the strut $(F+F_i)$ is zero, the blocks just touch the pin with no force in the springs. The pin is in contact with either the two blocks on the left or the two blocks on the right, depending on whether the joint is under tension or compressive loading.

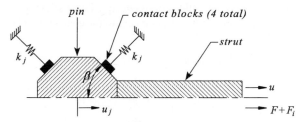

Figure 2. Simple model of a pinned joint.

As force F or F_i is increased, the displacements shown in Figure 3 occur, as shown by the dashed lines. The displacements create a normal force F_n and a frictional force F_f on the face of the contact blocks as illustrated in Figure 4. The pin moves distance u_j while the block on the right is compressed distance r and slides distance s.

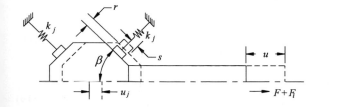

Figure 3. Illustration of displacements in the simplified pinned joint.

Figure 4. Forces on the contact block.

The relationships among u_j, r, and s are given by:

$$r = u_j \sin\beta \tag{11}$$

$$s = u_j \cos\beta \tag{12}$$

Summation of forces in the direction of the strut gives:

$$F + F_i = 2(F_n \sin\beta + F_f \cos\beta) \tag{13}$$

To simplify this derivation, we will assume the static and kinetic friction coefficients are equal for the pin/block interface. Using a friction coefficient μ, we can define:

$$F_f = \mu F_n = \mu k_j r \tag{14}$$

Combining the above three equations gives:

$$F + F_i = 2 k_j r (\sin\beta + \mu \cos\beta) \tag{15}$$

It is assumed that the blocks slide continuously without sticking. Sliding begins when the component of the force in the strut tangent to the contact interface exceeds the frictional force, or:

$$\frac{1}{2}(F + F_i) \cos\beta > \frac{1}{2}\mu(F + F_i) \sin\beta$$

$$\frac{1}{\mu} > \tan\beta$$

$$\beta < \tan^{-1}\left(\frac{1}{\mu}\right) \tag{16}$$

Using a typical value of $\mu=0.2$ in the above equation we obtain $\beta<78.7°$ for slipping to occur. Thus, this analysis limits β to the value given by Eq. (16).

Using Eq. (11), (12), and (15) we can solve for r, u_j, and s as:

$$r = \frac{F + F_i}{2 k_j (\sin\beta + \mu \cos\beta)} \tag{17}$$

$$u_j = \frac{r}{\sin\beta} = \frac{F + F_i}{2 k_j (\sin\beta + \mu \cos\beta) \sin\beta} \tag{18}$$

$$s = u_j \cos\beta = \frac{F + F_i}{2 k_j (\sin\beta + \mu \cos\beta) \tan\beta} \tag{19}$$

From the above relations we need to compute the energy loss during one cycle for a strut. If F varies harmonically, the energy dissipated by one joint during one cycle is obtained by multiplying the energy dissipated during a quarter cycle by four. Each of the two joints attached to the strut are assumed to have identical energy dissipation characteristics. Thus, the energy dissipation per strut is obtained by multiplying the energy dissipated per joint by two. At any time during a cycle, only two of the four blocks in each joint are in contact with the pin and creating friction losses. Therefore, the energy dissipated during one cycle is given by:

$$\Delta U = 8 \int_0^s 2 F_f \, ds = 16 \int_0^s \mu k_j r \, ds$$

$$= 16 \int_0^F \mu k_j \frac{F + F_i}{2 k_j(\sin\beta + \mu\cos\beta)} \frac{1}{2 k_j(\sin\beta + \mu\cos\beta)\tan\beta} dF$$

$$= \frac{2\mu (F + 2 F_i) F}{k_j(\sin\beta + \mu\cos\beta)^2 \tan\beta} \tag{20}$$

Eq. (21) gives the energy stored in the strut which includes two joints during one cycle using $F_i=0$. Typically, the vast majority of the energy is stored in the struts and energy storage in the joints could be neglected. Eq. (22) gives the loss factor for the strut.

$$U = \frac{F^2}{2 k_s} + 2 k_j r^2 = \frac{F^2}{2 k_j}\left(\frac{k_j}{k_s} + \frac{1}{(\sin\beta + \mu\cos\beta)^2} \right) \tag{21}$$

$$\eta = \frac{\Delta U}{2\pi U} = \frac{2\mu\left(1 + 2\dfrac{F_i}{F} \right)}{\pi\left(\dfrac{k_j}{k_s}(\sin\beta + \mu\cos\beta)^2 + 1 \right)\tan\beta} \tag{22}$$

Although Eq. (22) is based on several simplifying assumptions it does provide some interesting insights. Figure 5 illustrates the variation in loss factor as a function of the angle β assuming $k_j/k_s=500$ (a typical value) and $F_i=0$ (no initial load). The loss factor due to material damping in aluminum is typically $\eta \approx 0.001$. Thus, the above model would generally predict damping due to extensional friction to exceed material damping only for low values of the angle β. Since the pin is round, the actual damping may be related to a weighted integrated average of the curve illustrated in Figure 5. The integration would need to be weighted to account for the fact that the transmitted force

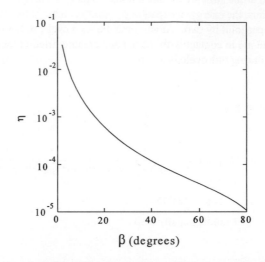

Figure 5. Illustration of predicted loss factor due to friction as a function of β.

(F) increases with angle β. This indicates that the extensional friction losses will generally be expected to be on the same order as material damping losses or less. The simplifying assumptions used in the above model may be inappropriate. Somewhat more elaborate models for joints with a clamping preload have been published.[5,13] These models draw essentially the same conclusion that extensional motion during contact does not appear to be a significant source of damping.

If deadband motion is possible and friction is present during the deadband motion, damping due to extensional motion can be large. For example, a shoulder bolt could be used as a pin in a joint and the bolt tightened to apply a fixed preload across the joint interface. If F_b is the bolt preload and δ is the deadband distance, the energy loss per cycle would be $2\mu F_b \delta$. This approach is generally not desirable because of joint wear and the slip-stick motion which would occur. The joints would not provide damping for low amplitude vibrations which could not overcome the frictional forces in the joint. However, it could be a significant source of damping for large amplitude vibrations.

4.2 Rotary Friction Damping Model

Consider the deformed and undeformed planar truss illustrated in Figure 6. The deformed geometry illustrates the joint rotations which will occur as the truss deforms. Structural stability requires at least one of the struts be rigidly attached to each joint (not pinned). Thus, the rotation angles shown in Figure 6 are measured relative to a reference strut which is assumed to be the rigidly attached strut at each joint. Note that for a given strut, the rotations at each end are not typically equal nor in the same direction.

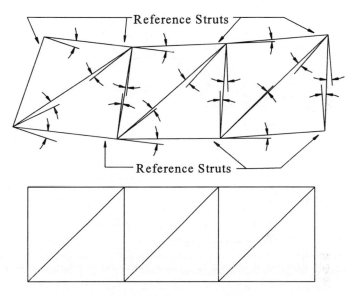

Figure 6.   Illustration of a undeformed (lower) and deformed (upper) three bay truss showing the expected rotations.

The derivation of an expression for friction losses due to rotary displacements is based on the strut model illustrated in Figure 7 which assumes the strut has a pinned joint at both ends. A strut of length L and pin radius r has an initial load, F_i, in the strut. The pins are assumed to be rigidly attached to the strut. The struts are represented by a simple line while the pin is represented by a large circle for clarity. The box around the pin represents the joint which holds the pin in place. The load F_i could be due to gravity or strut fabrication errors (making them too long or too short). The truss to which the strut is connected undergoes sinusoidal displacements causing load F to be added to the strut and causing rotations of θ_1 and θ_2 at joint #1 and #2, respectively. F, θ_1, and θ_2 have a sinusoidal variation with time (t) and are assumed to be linearly related as follows:

$$F = F_* \sin(\omega t)$$

$$\theta_1 = \theta_{1*} \sin(\omega t)$$

$$\theta_2 = \theta_{2*} \sin(\omega t)$$

$$F_* = c_1 \theta_{1*} = c_2 \theta_{2*} \tag{23}$$

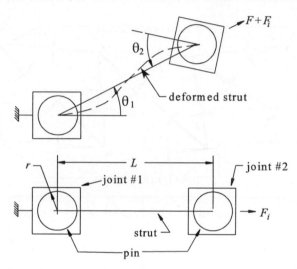

Figure 7.   Illustration of a model for friction losses due to rotary displacements.

The parameters c_1 and c_2 are constants determined for each strut for a given vibration mode. The cylindrical pins in the ends of the strut transmit the loads in the struts to the joints as illustrated in Figure 8. Because of the presence of a normal force (F_n) across the joint, a frictional force (F_f) will be created as the joint tries to rotate. The force F_f will cause a moment in the ends of the strut and a very slight bending of the strut, illustrated by the dashed line in Figures 7 and 8.

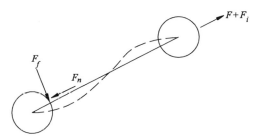

Figure 8. Illustration of normal and friction forces in a pinned joint.

A "stick and slip" type of motion is assumed to occur. That is, during portions of each cycle, the joint sticks to the strut, preventing rotary slip. Figure 9 represents peaks in the oscillatory motion where each of the joints undergo rotations of $\pm\theta_{1*}$ and $\pm\theta_{2*}$. Note that the friction induced moment acting on the ends of the strut creates rotations at the ends and thereby reduces the rotational slip. For example, the rotational slip of joint #1 is reduced by θ_{1+f} at $\theta_1 = +\theta_{1*}$ and by θ_{1-f} at $\theta_1 = -\theta_{1*}$. Note that as long as $F_i \neq 0$, then $\theta_{1+f} \neq \theta_{1-f}$ because the normal force (F_n) has different magnitudes at the displacement peaks. The net rotational slip of joint #1 during one half of a cycle is given by $2\theta_{1*} - \theta_{1+f} - \theta_{1-f}$. If $2\theta_{1*} < \theta_{1+f} + \theta_{1-f}$ then no slipping occurs (the pin sticks to the joint) and damping due to macroslip does not occur. Similar arguments apply to joint #2.

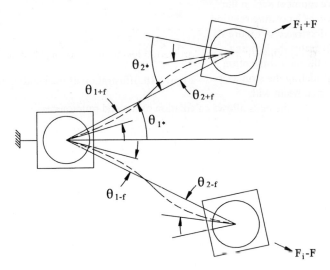

Figure 9. Illustration of rotary motions in a strut with pinned joints.

The magnitudes of θ_{1+f}, θ_{1-f}, θ_{2+f}, and θ_{2-f} need to be evaluated. Assuming that rotational slip occurs at both joints, the normal and frictional forces (F_n and F_f) transmitted to the pin are illustrated in Figure 8 and are given by:

$$F_f = \mu F_n = \mu (F_i + F) \qquad (24)$$

where μ is the kinetic friction coefficient. The moments applied to the strut at joints #1 and #2 due to friction are given by:

$$M_{1f} = -r\,|F_f|\,\text{sign}\!\left(\frac{\partial\theta_1}{\partial t}\right) = -\mu r\,|F_i + F|\,\text{sign}\!\left(\frac{\partial\theta_1}{\partial t}\right) \qquad (25)$$

$$M_{2f} = -\mu r\,|F_i + F|\,\text{sign}\!\left(\frac{\partial\theta_2}{\partial t}\right) \qquad (26)$$

where r is the radius of the pin and the sign function returns the sign of the angular velocity. Note that $M_{1f} = \pm M_{2f}$ if rotational slip occurs and the joints have identical properties. The sign convention used here is that a positive moment acts in the same direction as a positive rotation (θ_1 or θ_2) as illustrated in Figure 7. Figure 10 illustrates the bending directions for the sign convention used here. Modeling the strut as a simply supported beam with

Figure 10. Illustration of the bending directions.

moments applied at the ends allows calculation of the end rotations as:

$$\theta_{1+f} = \frac{\left(M_f\right)_{F=+F_*} L}{bEI} = \frac{\mu r L F_*}{bEI}\left|\frac{F_i}{F_*}+1\right| \qquad (27)$$

$$\theta_{1-f} = \frac{\left(M_f\right)_{F=-F_*} L}{bEI} = \frac{\mu r L F_*}{bEI}\left|\frac{F_i}{F_*}-1\right| \qquad (28)$$

where:

E = Young's modulus of the strut
I = area moment of inertia of the strut
b = 2, if $M_{1f} = -M_{2f}$
= 12, if $M_{1f} = M_{2f}$.

It is noted that Eq. (27) and (28) assume that both joints undergo rotational slip. If rotary slip does not occur at both joints (e.g., if one end of a strut is rigidly attached to a joint or does not undergo macroslip), then the above relationships need to be modified.

The energy dissipated by a joint during one cycle can be computed by integrating the energy dissipated over one half of a cycle and multiplying the result by 2. Eq. (29) gives the energy loss for joint #1.

$$\Delta U_1 = 2 \int_{-\theta_{1*}+\theta_{1-f}}^{\theta_{1*}-\theta_{1+f}} M_{1f} d\theta_1 = 2\mu r \int_{-\theta_{1*}+\theta_{1-f}}^{\theta_{1*}-\theta_{1+f}} |F_i + c_1\theta_1| \, d\theta_1$$

$$= \mu r \left[\left(\theta_{1*} - \theta_{1+f} + \frac{F_i}{c_1} \right) \left| F_i + c_1(\theta_{1*} - \theta_{1+f}) \right| \right.$$

$$\left. - \left(-\theta_{1*} + \theta_{1-f} + \frac{F_i}{c_1} \right) \left| F_i + c_1(-\theta_{1*} + \theta_{1-f}) \right| \right] \tag{29}$$

The energy loss for joint #2 and the total loss for the strut are given by:

$$\Delta U_2 = \mu r \left[\left(\theta_{2*} - \theta_{2+f} + \frac{F_i}{c_2} \right) \left| F_i + c_2(\theta_{2*} - \theta_{2+f}) \right| \right.$$

$$\left. - \left(-\theta_{2*} + \theta_{2-f} + \frac{F_i}{c_2} \right) \left| F_i + c_2(-\theta_{2*} + \theta_{2-f}) \right| \right] \tag{30}$$

$$\Delta U = \Delta U_1 + \Delta U_2 \tag{31}$$

It is assumed that the vast majority of the strain energy of the truss is attributed to axial tension or compression forces acting on each strut. Neglecting the energy stored in the joints and the bending energy due to the friction induced moments at the strut ends, the energy stored in the strut during a cycle is given by:

$$U = \frac{F_*^2}{2k_s} \tag{32}$$

where k_s is the strut stiffness. The loss factor for rotary friction losses is given as:

$$\eta = \frac{\Delta U}{2\pi U}$$

$$= \frac{\mu r k_s}{\pi} \left[\frac{1}{c_1} \left(1 - \frac{c_1 \theta_{1+f}}{F_*} + \frac{F_i}{F_*} \right) \left| \frac{F_i}{F_*} + 1 - \frac{c_1 \theta_{1+f}}{F_*} \right| \right.$$

$$+ \frac{1}{c_1} \left(1 - \frac{c_1 \theta_{1-f}}{F_*} - \frac{F_i}{F_*} \right) \left| \frac{F_i}{F_*} - 1 + \frac{c_1 \theta_{1-f}}{F_*} \right|$$

$$+ \frac{1}{c_2} \left(1 - \frac{c_2 \theta_{2+f}}{F_*} + \frac{F_i}{F_*} \right) \left| \frac{F_i}{F_*} + 1 - \frac{c_2 \theta_{2+f}}{F_*} \right|$$

$$+ \left. \frac{1}{c_2} \left(1 - \frac{c_2 \theta_{2-f}}{F_*} - \frac{F_i}{F_*} \right) \left| \frac{F_i}{F_*} - 1 + \frac{c_2 \theta_{2-f}}{F_*} \right| \right] \qquad (33)$$

If initial strut loads are neglected ($F_i = 0$) and the joints do not "stick" due to friction ($\theta_{1+f} = \theta_{1-f} = \theta_{2+f} = \theta_{2-f} = 0$), then Eq. (31) reduces to:

$$\eta = \frac{2\mu r k_s}{\pi} \left(\frac{1}{c_1} + \frac{1}{c_2} \right) \qquad (34)$$

or using Eq. (23) we get:

$$\eta = \frac{2\mu}{\pi} \frac{r k_s}{F_*} (\theta_{1*} + \theta_{2*}) \qquad (35)$$

Eq. (35) agrees with a relationship published by Hertz and Crawley.[14]

Examining Eq. (33), (34) and (35) provides some interesting insights. Using the typical values of $\mu = .2$, $r k_s / F_* = 50$, and $\theta_{1*} = \theta_{2*} = 0.001$, Eq. (35) predicts a loss factor of $\eta \approx 0.01$, which is about 10 times larger than typical values for material damping. Thus, significant amounts of damping are available when loaded members are rotated. If displacements are small, c_1 and c_2 should be constant and Eq. (34) would predict damping which is independent of amplitude (i.e., it is no longer dependent on F_*, θ_{1*} or θ_{2*}). This is disturbing since measured damping data from truss structures typically shows a decrease in damping as amplitude decreases. However, note that Eq. (33) retains amplitude dependence, particularly through the terms θ_{1+f}, θ_{1-f}, θ_{2+f}, and θ_{2-f}. By using

Eq. (23), the c_1 and c_2 parameters can be removed as shown in Eq. (36).

$$
\eta = \frac{\mu \, r \, k_s}{\pi F_*} \left[\, \theta_{1*} \left(1 - \frac{\theta_{1+f}}{\theta_{1*}} + \frac{F_i}{F_*} \right) \left| \frac{F_i}{F_*} + 1 - \frac{\theta_{1+f}}{\theta_{1*}} \right| \right.
$$

$$
+ \, \theta_{1*} \left(1 - \frac{\theta_{1-f}}{\theta_{1*}} - \frac{F_i}{F_*} \right) \left| \frac{F_i}{F_*} - 1 + \frac{\theta_{1-f}}{\theta_{1*}} \right|
$$

$$
+ \, \theta_{2*} \left(1 - \frac{\theta_{2+f}}{\theta_{2*}} + \frac{F_i}{F_*} \right) \left| \frac{F_i}{F_*} + 1 - \frac{\theta_{2+f}}{\theta_{2*}} \right|
$$

$$
+ \, \theta_{2*} \left(1 - \frac{\theta_{2-f}}{\theta_{2*}} - \frac{F_i}{F_*} \right) \left| \frac{F_i}{F_*} - 1 + \frac{\theta_{2-f}}{\theta_{2*}} \right| \, \right] \qquad \textbf{(36)}
$$

During the decay of a vibration mode, initial strut loads (F_i) can become larger than the cyclic loads (F_*) causing the strut to "stick" for a greater portion of each cycle, thereby reducing the damping. Furthermore, as gravity loads increase (thereby increasing F_i), Eq. (36) will predict a decrease in damping. Finally, although Eq. (36) appears quite attractive, applying it to real structures can be difficult since it is dependent on accurate knowledge of initial strut loads, friction coefficients, and the amount of "stuck rotation." These data are often not readily available. Thus, Eq. (36) probably has more qualitative than quantitative value. Nevertheless, rotational friction can be a significant source of damping.

4.3 Impact Damping Model

If a pinned joint is free to rotate, a small amount of clearance around the pin is usually needed. As load reversal occurs in a joint, a relative movement between the strut and the joint can occur due to this clearance. This relative motion allows impacting of the contacting surfaces. Figure 11 illustrates a

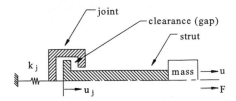

Figure 11. Illustration of a simple model of a joint with a gap or deadband.

simple model of a joint with clearance (deadband). The contribution of impacting to the loss factor for a strut is difficult to obtain. If we assume the joint is fixed, the energy loss in an inelastic collision is approximated by:

$$\Delta U = \frac{1}{2} M v_r^2 (1 - e^2)$$ (37)

where:

v_r = relative velocity of the impacting surfaces

M = "effective" mass of the truss structure

e = coefficient of restitution

The coefficient of restitution is a function of the material properties of the contacting surfaces as well as their shape.

One of the difficulties in evaluating Eq. (37) is the proper "effective" mass to use in a truss structure composed of many elements. Gronet et. al.[13] lists the following equation for estimating the loss factor in truss structures:

$$\eta = \frac{1}{2\pi} \left[\frac{k_s}{k_j} \right]^2 (1 - e^2)$$ (38)

where k_s is the stiffness of the strut and k_j is the stiffness of the strut joint. The assumptions made to derive Eq. (38) are not given. Assuming reasonable values of e=0.9 and k_s/k_j=0.01 gives η=3 x 10^{-6}. Eq. (38) predicts significant amounts of damping only when the joint stiffness is comparable to the tube stiffness, which is unreasonable for most truss structures. It is suspected that Eq. (38) is too simplistic to represent this complicated energy loss mechanism. Furthermore, the problem of assessing the coefficient of restitution still remains.

The coefficient of restitution is a function of the materials which come into contact and the geometry of the contacting surfaces. Lankarani and Nikravesh[15] describe how to relate a hysteresis damping coefficient to the coefficient of restitution of a sphere driven by a linkage and contacting a second fixed sphere. The hysteresis damping coefficient can then be used to predict energy dissipation during the contact period. The relations derived apply to "external" contact with a Hertz based model for elasticity of the contact surfaces. The linkage is considered as rigid. A pinned joint would be described as an "internal" contact. In internal contact problems with low relative velocities, the stiffness and mass of structure surrounding a joint can dramatically influence the structural response.

The response of the structure surrounding a joint can be an additional source of energy dissipation during impact. When a pinned joint traverses the deadband zone, the impact which occurs at the end of that cycle transmits a step impulse to the rest of the truss structure. That step impulse can excite higher frequency vibration modes. Thus, vibration energy is transferred from the low frequency global truss modes to the high frequency truss modes where it is dissipated rapidly by material and friction damping mechanisms. This form of energy dissipation in structures using pinned joints has been reported by others[16,17].

One popular method of characterizing the dynamic properties of a strut is the force-state-mapping technique[18,19,20,21]. Sections of a truss or individual struts are subjected to dynamic loading and resulting displacements, accelerations, and applied forces are measured. From these data, a map of the force-displacement-velocity domain is obtained. Analytical models of strut behavior could then be fit to the surface which includes the effects of nonlinear stiffness and damping. These analytical models of truss components can then be extended to create an analytical model of the entire truss.

The magnitude of damping due to impact in a force-state-mapping test can be significantly influenced by the nature of the test fixture attached to a joint. This may influence the accuracy of a force-state mapping procedure for determination of joint damping characteristics. The fixture might either encourage or discourage the rate of energy dissipation through impacting, as compared with behavior in a truss. Ferney and Folkman[22] report results of force state mapping tests completed with pinned joints. These tests showed multiple impacts occurs during each load cycle in the force-state-map tests. However, when mass was added to the test article, the number of rebounds and the magnitude were modified. Analytical models of a strut easily confirm this behavior and indicate that with a soft surrounding structure with sufficient mass, the rebounding can be eliminated and a "soft" impact is obtained. That is, by modifying the stiffness and/or mass of the structure surrounding a joint, the dynamic behavior and damping of that joint is modified.

It is difficult to accurately predict the damping due to impact at this time. There are multiple mechanisms at work and they are difficult to predict. It is further complicated by the fact that both friction and impact damping mechanisms occur together. More research in this area is needed.

5. Measured Data

Utah State University is currently developing an experiment titled Joint Damping Experiment (JDX) to fly on the Space Shuttle as a Get Away Special payload. This project is funded by NASA's IN-Space Technology Experiments Program and is scheduled to fly in July 1995. The objectives of JDX include development of a small-scale shuttle flight experiment which allows researchers to characterize the influence of gravity and joint gaps on structural damping and dynamic behavior of a small-scale truss and application of nonlinear finite element modeling capability to simulate the measured behavior. This will allow a better understanding and/or prediction of the structural dynamics occurring in a pin-jointed truss.

A photograph of the JDX truss is given in Figure 12. The bottom of the truss is attached to a base plate that provides a cantilevered boundary condition for that end of the truss. A rigid plate is attached to the top bay of the truss to provide a tip mass which lowers the natural frequencies. The excitation system can preferentially excite three modes in the truss; two bending modes and a torsional mode. The two bending modes are the two lowest frequency modes of the truss. These two bending modes are described as "bend 1" and "bend 2" modes. The lacing of the struts separates the two bending mode frequencies so each can be easily identified. The torsional mode consists of a rotational motion about the long axis of the truss. The small plates attached to the tip

Figure 12. Photograph of the JDX truss.

mass in Figure 1 are used by the excitation system. The damping and dynamic characteristics are observed by recording the free decay of the modes.

The strut and joint design for an "unlocked" joint is illustrated in Figure 13. The aluminum struts can be adjusted in length by turning the threaded strut tubes. The joints utilize press-fit hardened steel inserts to minimize joint wear. Hardened steel shoulder bolts are used for joint pins. The gap in the joints can be adjusted by using different diameter shoulder bolts. The JDX truss also uses "locked" joints which do not permit any deadband motion or rotations. The locked joint design is similar to Figure 13 except no inserts are used and the shoulder bolts are oversized

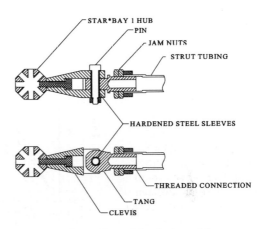

Figure 13. Illustration of the design of the struts.

such that they must be press fit into the joint. Additionally, shims are installed between the clevis and tang pieces and the shoulder bolt is tightened to apply a high preload across the joint interface. Impacting and macro-slip damping mechanisms are not present in the locked joints. By locking only a portion of the joints in the truss, a method of tailoring truss dynamics is obtained.

The initial tests of the JDX truss showed that with all the joints unlocked, the vibration modes were almost obscured with high frequency hash. The high frequency hash was a caused by impacting of the numerous truss joint gaps. It was clearly shown that only a few unlocked joints in a truss would dramatically influence the dynamic behavior. The flight model configuration of the truss has only eight unlocked joints located at the top of the first bay as illustrated in Figure 14.

The nominal radial clearance

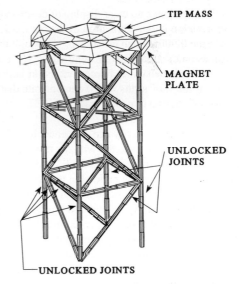

Figure 14. Location of unlocked joints.

between the pins and the strut inserts in the JDX flight model truss design was 0.0003 inch. This represents a very close fit for a 0.1875 inch pin. This is essentially the largest pin which could be freely inserted into the joint. Figure 15 illustrates quasi-static pull tests of a JDX strut which has a pinned joint on one end and a locked joint on the other. The hysteresis behavior of the strut with two different pin diameters are shown in Figure 15. The 0.00025 inch radial pin gap represents the class of fit used in the flight model truss. The quasi-static tests using the 0.00025 inch

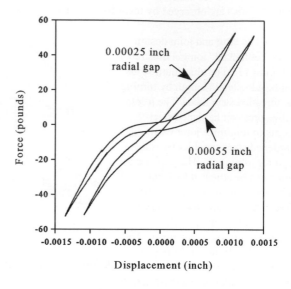

Figure 15. Hysteresis in quasi-static pull tests of a JDX strut.

radial gap shows no clear display of deadband, but significant hysteresis due to friction. Imprecision in the fabrication of the pin masks the deadband in these tests, as compared with the 0.00055 inch radial gap. Although the gap in the joint is small, the dynamic effects on the truss remain significant. The quasi-static tests in Figure 15 do not reflect the dynamic behavior. Ferney and Folkman[22] reported an initial series of force-state-map tests of the JDX struts which demonstrate that the joint gaps give rise to multiple rebounds during dynamic loading.

Because gravity can influence the behavior of a pinned joint, ground tests of the JDX truss were conducted with the truss in two orientations with respect to the gravity vector. Figure 16 illustrates the 0° and 90° truss orientations. The 0° orientation minimizes the gravity preload in the struts. Although the preload in the truss is minimized in the 0° orientation, it is not zero since the joints must still support the weight of the truss. The 90° orientation provides the maximum gravity induced strut preloads. It is important that the truss was carefully assembled such the preloads due imprecise strut lengths are virtually eliminated. That is, if gravity loads are removed, the deadband region is easily traversed. This turns out to be a difficult procedure in that the strut lengths must be set near the same level of accuracy as the joint pin gaps. In practice, this is difficult to obtain.

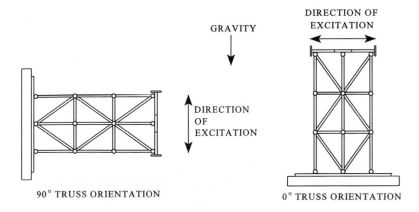

Figure 16. Illustration of 0° and 90° test orientations.

Ground tests of the flight model JDX tests are illustrated in Figures 17, 18, 19, and 20. These figures are plots of accelerations of the tip mass in the direction of the excitation. The free decay of the bend 1 mode with all joints locked is illustrated in Figure 17. Low damping and a typical exponential decay is observed. Figure 18 shows the initial decay over a 1 second time period. Because the JDX truss is displaced and released from rest, a there is a small amount of higher mode content which quickly damps out. Figure 19 illustrates the decay of the bend 1 mode for a truss in the 0° orientation with eight unlocked joints. Note the rapid decay and the large amount of high frequency hash in the decay. Also the frequency is about 10% lower when compared with the truss will all locked joints. Figure 20 illustrates the decay of the bend 1 mode with the truss in the 90° orientation and with eight unlocked joints. Note that damping is decreased relative to the 0° orientation test but the damping is still much greater than that of a locked truss. Note how the initial decay is not symmetric about the equilibrium position. The accelerometers used are Kister K-Beams which produce a DC output and since the accelerometer is oriented in the direction of the gravity vector, it reads approximately 1 g at equilibrium. An acceleration of 0 g's coincides with the position where the joints supporting the weight of the truss would traverse their deadband region. Clearly, when the highly loaded joints traverse the deadband region, the dynamics of the truss are greatly modified. The damping is reduced significantly when the amplitude of the peak accelerations drops below 1 g since the active joints remain under preload at all times. At small amplitudes, the damping approaches that of the truss with all joints locked. The bend 2 mode recorded similar behavior.

The results reported above are similar to those reported earlier for the JDX engineering model truss[23]. Damping of the truss can be inferred from the logarithmic decrement of the decay if we treat the truss as a single degree of freedom system and assuming energy is not transferred to other modes. Reference 23 reports the logarithmic decrement of the bend 1 mode for the locked, 0°, and 90° truss orientations as a function of acceleration amplitude. It was reported that:

1) damping rates can change by a factor of 2 to 5 as a result of simply changing the orientation of a truss;
2) the addition of a few unlocked joints to a truss structure can increase the damping by a factor as high as 40;
3) damping is amplitude dependent;
4) at low amplitudes the damping in the 90° orientation approaches that of the locked truss.

At low amplitudes, strut preloads may discourage a macroslip mode of friction damping as well as impacting. In this case, one would expect the damping to approach that of a locked joint. Although the damping becomes smaller at low amplitudes in the 0° truss orientation with unlocked joints, it is still significantly higher than the truss with locked joints.

The high frequency hash in Figures 19 and 20 is suspected to be caused by impacts occurring in the joints. This is more clearly shown in Figure 21 which shows the first 0.6 seconds of the decay in Figure 20 for the 90° truss orientation. Figure 21 shows that the occurrence of the high frequency hash corresponds to the traversal of the deadband region. After approximately 0.4 seconds, the highly loaded joints can no longer traverse the deadband region and the high frequency hash becomes greatly diminished.

Although the test results indicated that impacting is a significant source of damping in pinned joints, friction is also suspected to provide a significant source of energy dissipation. The joints which carry the majority of the load in the 90° orientation tests with unlocked joints should not be traversing the deadband zone after the amplitude drops below 1 g. Lightly loaded joints in the truss could still be traversing the deadband zone, but the magnitude of the impacts ought to be small. Thus, much of the damping from the 90° tests with unlocked joints is suspected to be predominately from a friction mechanism.

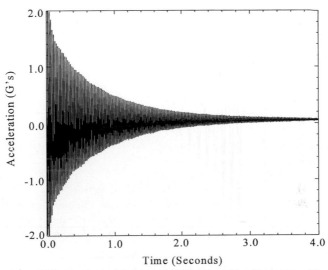

Figure 17. Decay of the bend 1 mode with the truss in a 0° orientation with all joints locked.

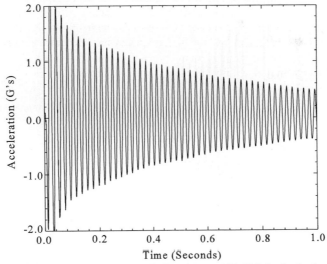

Figure 18. Initial decay of the bend 1 mode for a truss will all joint locked.

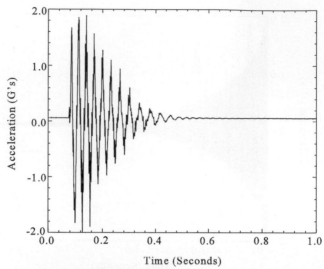

Figure 19. Decay of the bend 1 mode with the truss in a 0° orientation and 8 unlocked joints.

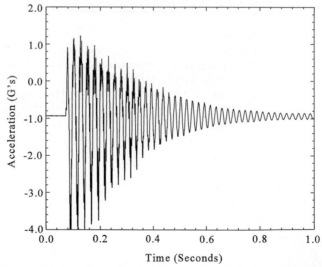

Figure 20. Decay of the bend 1 mode with the truss in a 90° orientation and 8 unlocked joints.

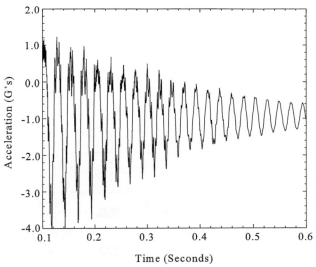

Time (Seconds)

Figure 21. Initial decay of the truss in the 90 orientation with eight unlocked joints.

6. Nonlinear Finite Element Model of a Joint

An effort is made here to describe a simple nonlinear finite element model of a joint which has clearance fit pins. The objective is to try to keep the model as simple as possible and yet capture the most important features of the joint's behavior. Figure 22 illustrates the model. The coordinate system in Figure 22 is aligned such that the x axis is aligned with the strut and the y and z axes are orthogonal to the x axis. Elements 1 and 2 represent beam elements used to model the structural portion of the clevis and tang, respectively. Elements 3 and 4 are gap elements used to model the deadband in the truss. These gap elements provide a very high stiffness when the gap is closed and a very small stiffness when it is open. Two gap elements are needed; one which closes when the strut is in tension and the other closes when the strut is in compression. The gap elements also support sliding friction when the gap is closed. Note that nodes 5 and 6 for these two elements are to be offset in the x direction by the radius of the pin. Thus if the pin rotates while the gap is closed, friction in the gap elements can provide damping. Also, the normal force across these gap elements (and thus the friction force as well) changes as the load on the strut changes. Elements 5 and 6 are also gap elements. These elements have a fixed preload across them and are closed at all times. Elements 5 and 6 provide a friction force during the deadband region. Note that nodes 7, 8, 9, and 10 are offset in the y or z directions by the radius of the pin so that elements 5 and 6 provide friction to resist both translations and rotations. Element 7 is a viscous damping element to provide an equivalent viscous damping parameter for the joint. Element 8 is a multipoint constraint element. Element 8 forces the y and z translational degrees of freedom of nodes 2 and 3 to be identical. Thus, nodes 2 and 3 are a hinge point. Note that in general, nodes 2 and 3 would be coincident and rigid elements would be needed to connect node 2 to nodes 5, 7, and 8 and to connect node 3 to nodes 6, 9, and 10.

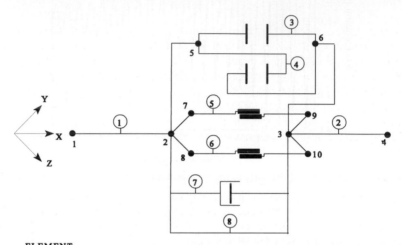

ELEMENT
NUMBER DESCRIPTION

1, 2   BEAM ELEMENTS
3, 4   GAP ELEMENTS TO SIMULATE DEADBAND AND FRICTION
5, 6   GAP ELEMENTS TO SIMULATE FRICTION
7   VISCOUS DAMPING ELEMENT
8   MULTIPOINT CONSTRAINT

Figure 22. Simple finite element model of a pinned joint.

The above model is simplistic. It can capture the quasi-static behavior illustrated in Figure 15. It can also examine driving of higher modes as the gaps close. Nevertheless, selection of friction coefficients, gap element preloads, and viscous damping coefficients is not a simple process. Furthermore, even though this model is simple, difficult convergence problems can occur if the truss being modeled has a large number of pinned joints. Much work in this area remains to be done.

7. Conclusions

If a truss structure utilizes a pinned joint with clearance fit pins, the joints can become the primary source of damping in the structure. The resonant frequency and driving of higher modes are also significantly influenced. The mechanisms of joint damping are not well understood. A review of literature related to joint damping is presented.

Expressions for damping due to friction caused by either extensional or rotary motions were derived. If deadband is neglected, damping due to friction is small for extensional motions of a strut. A significant source of damping can be caused by extensional motions if friction forces are present during deadband motion. Rotary friction can be a significant source of damping. An expression is derived for damping due to rotary motion which possesses amplitude dependence. Friction damping should be usually become small as the amplitude of oscillations gets small. A slick-slip motion should gradually terminate and damping shifts from a macroslip mode to a microslip mode.

Measured data for a three bay truss containing eight pinned joints was presented. The measured data clearly shows that as gravity induced preloads are decreased, damping increases. Impacting is demonstrated by the observation of high frequency hash in the decay data. Driving of higher modes by impacting is suspected to be a significant source of damping. High preloads in the 90° truss orientation can prevent impacting in the joints carrying the majority of the loads (as long as the amplitude of the oscillations is smaller than 1 g). Damping in this time period should be dominated by friction losses. Measured data confirms that friction damping is a significant factor.

A simple nonlinear finite element model of a pinned joint is presented. The model is simplistic. Because nonlinear solutions can require large solution times and gap elements can have severe convergence problems, a simple model that captures the most important joint features would be advised as a beginning point. Determination of friction and equivalent viscous damping parameters for the nonlinear finite element model is difficult. Much work in this area remains to be completed.

8. Acknowledgment

This work was partially supported by the NASA INSTEP program, funded through NASA Langley Research Center (LaRC) under contract NAS1-19418. The support of Mark Lake at LaRC as technical monitor is gratefully acknowledged.

9. References

1.   J. M. Hedgepeth and L. R. Adams, "*Design Concepts for Large Reflector Antenna Structures*," NASA Contractor Report 3663, Jan. 1953, p. 73.

2.   H. Ashley, "On Passive Damping Mechanisms in Large Space Structures," *AIAA Journal of Guidance, Control and Dynamics*, **Vol. 21**, No. 5, 1984, p. 448-455.

3.   R. Plunkett, "Friction Damping," *Damping Applications for Vibration Control*, AMD **Vol. 38**, edited by P. J. Torvik, ASME, N. Y., Nov. 1980, pp. 65-74.

4.   S. L. Folkman and F. J. Redd, "Gravity Effects on Damping of a Space Structure with Pinned Joints," *AIAA Journal of Guidance, Control, and Dynamics*, **Vol. 13**, No. 2, March-April 1990, p. 228.

5.   E. F. Crawley, et al, "*Prediction and Measurement of Damping in Hybrid Scaled Space Structure Models*," **Report SSL 7-88**, Space Systems Laboratory, Dept. of Aeronautics and Astronautics, MIT, July 1988.

6.   J. P. Den Hartog, *Mechanical Vibrations*, McGraw-Hill, New York, 4 ed., 1956, p. 362.

7.   C. F. Beards and J. L. Williams, "The Damping of Structural Vibration by Rotational Slip in Joints," *Journal of Sound and Vibration*, **Vol. 53**, No. 3, 1977, pp. 333-340.

8.   S. Dubowsky and F. Freudenstein, "Dynamic Analysis of Mechanical Systems with Clearances, Part 1: Formation of the Dynamic Model," *ASME Journal of Engineering for Industry*, February 1971, p. 305-309.

9.   S. Dubowsky and F. Freudenstein, "Dynamic Analysis of Mechanical Systems with Clearances, Part 2: Dynamic Response," *ASME Journal of Engineering for Industry*, February 1971, p. 310-316.

10.   S. Dubowsky, "On Predicting the Dynamic Effects of Clearances in Planar Mechanisms," *ASME Journal of Engineering for Industry*, February 1974, p. 317-323.

11.   A. A. Ferri, "Modelling and Analysis of Nonlinear Sleeve Joints of Large Space Structures," *AIAA Journal of Spacecraft and Rockets*, **Vol. 25**, No. 5, Sept.-Oct. 1988, p. 354-365.

12.   A. Bindemann and A. A. Ferri, "The Influence of Alternate Friction Laws on the Passive Damping of a Flexible Structure," *Proceedings of the 36th AIAA/ASME/ASCE/AHS/ASC Structures, Structural Dynamics and Materials Conference*, April 1995, AIAA-95-1178.

13.   M. J. Gronet, E. D. Pinson, H. L. Voqui, E. F. Crawley, and M. R. Everman, "*Preliminary Design, Analysis and Costing of a Dynamic Scale Model of the NASA Space Station*," NASA Contractor Report 4068, July 1987, pp. 13-31.

14.   T. J. Hertz, E. F. Crawley, "Displacement Dependent Friction in Space Structure Joints," *AIAA Journal*, **Vol. 23**, No. 12, Dec. 1985, p. 1998-2000.

15.   H. M. Lankarani and P. E. Nikravesh, "A Contact Force Model with Hysteresis Damping for Impact Analysis of Multibody Systems," *ASME Journal of Mechanical Design*, Sep. 1990, **Vol. 112**, p.369.

16.   J. Onoda, T. Sano, and K. Minesugi, "Passive Vibration Suppression of Truss by Using Backlash," *The 34th AIAA/ASME/ASCE/AHS/ASC Structures, Structural Dynamics, and Materials Conference*, LaJolla Ca., April 19-22, 1993, Paper Number AIAA-93-1549-CP.

17.   Y. Hayasaka, et. al., "Analysis of Nonlinear Vibration of Space Apparatuses Connected with Pin-Joints," *43rd Congress of the International Astronautical Federation*, Washington D.C., Sep. 1-4, 1992, Paper Number IAF-92-0315.

18.   E. F. Crawley and A. C. Aubert, "Identification of Nonlinear Structural Elements by Force-State Mapping," *AIAA Journal*, **Vol. 24**, Jan. 1986, pp. 155-162.

19.   E. F. Crawley and K. J. O'Donnell, "Force-State Mapping Identification of Nonlinear Joints," *AIAA Journal*, **Vol. 25**, July 1987, pp. 1003-1010.

20.   B. P. Masters, E. F. Crawley, and M. C. van Schoor, "Global Structure Modeling Using Force-State Component Identification," *Proceedings of the 35th AIAA/ASME/ASCE/AHS/ASC Structures, Structural Dynamics and Materials Conference*, April 1994, AIAA-94-1519.

21.   S. J. Bullock and L. D. Peterson, "Identification of Nonlinear Micron-Level Mechanics for a Precision Deployable Joint," *Proceedings of the 35th AIAA/ASME/ASCE/AHS/ASC Structures, Structural Dynamics and Materials Conference*, April 1994, AIAA-94-1347.

22.   B. D. Ferney and S. L. Folkman, "Results of Force-state Mapping Tests to

Characterize Struts Using Pinned Joints," *Proceedings of the 36th AIAA/ASME/ASCE/AHS/ASC Structures, Structural Dynamics and Materials Conference*, April 1995, AIAA-95-1150.

23.   S. L. Folkman, E. A. Rowsell and G. D. Ferney, "Gravity Effects on Damping of a Truss Using Pinned Joints," *Proceedings of the 35th AIAA Dynamics Specialists Conference*, April 1994, AIAA-94-1685.

Dynamics with Friction: Modeling, Analysis and Experiment, pp. 169–195
edited by A. Guran, F. Pfeiffer and K. Popp
Series on Stability, Vibration and Control of Systems Series B: Vol. 7
© World Scientific Publishing Company

DESIGN OF A FRICTION DAMPER
TO CONTROL VIBRATION OF TURBINE BLADES

JHY-HORNG WANG
Department of Power Mechanical Engineering
National Tsing Hua University
Hsin Chu, Taiwan, R.O.C.

ABSTRACT

Friction damping can be found everywhere in daily life, yet it is difficult to design an effective, practical friction damper. One of the few successful examples of a practical friction damper is the friction damper used in turbine blades. In this chapter, the damper used to control the vibration of blades is adopted as an example to discuss the key points involved in designing a friction damper. First, the theoretical model of friction interfaces is presented, and the verification of the model by experiment is described. The friction damper used in quasi-solid blades is then used as an example to illustrate in detail the process of designing a friction damper.

1. Introduction

Friction can dissipate energy, and the friction damper is a typical device used to reduce the vibration of a structure. Theoretically, a friction damper is a powerful device to control the vibration of a structure; in practice, however, friction dampers have been used successfully in only a few types of applications. One such successful application is the friction damper used in turbine blades.

Two types of construction have been used in turbine blades: free-standing blades, and grouped (or packet) blades. In the free-standing blade design, each blade stands alone; in the grouped-blade design, several blades are grouped together into a packet using a shroud or tie-wire. A blade-to-ground friction damper can be used with free standing blades to control the blade vibration, as shown schematically in Fig. 1. In the grouped-blade design, the shroud interfaces and the interfaces between the tie-wire and blades provide a so-called blade-to-blade friction damper, as shown schematically in Figs. 2 and 3, respectively.

Both the blade-to-blade and blade-to-ground dampers can dissipate vibration energy only when relative motion (slipping) occurs between the friction interfaces. Hence, the modeling and analysis of the relative motion between the friction interfaces is the key problem in designing a friction damper. The frictional characteristics of the friction interfaces in turbine blades fall into the category of dry friction. There are two theoretical approaches to model dry friction interfaces: the macroslip and microslip approaches. In the microslip approach[1,2], a relatively detailed analysis of the friction interfaces must be carried out. Generally speaking, the microslip approach can provide more accurate results only when the normal preload between the interfaces is very high,

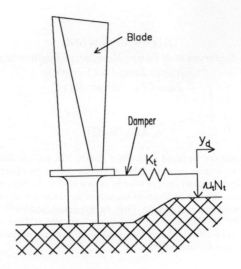

Fig. 1   Single blade with blade-to-ground damper

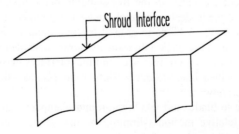

Fig. 2   Groupped blade with integral shroud

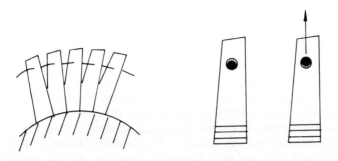

Fig. 3   Groupped blade with loosing tie-wire which is brought into contact with the blade via centrifugal force

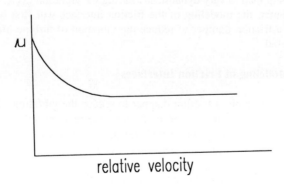

relative velocity

Fig. 4   Friction coefficient as a function of relative velocity of contact interface

and at the expense of higher computational effort. In the macroslip approach[3-5], the entire surface of the friction interface is assumed to be either slipping or stuck. This simple assumption is one reason for the widespread popularity of the macroslip approach; a second reason is that the friction interfaces can dissipate energy efficiently only when apparent macroslip and stick occur. In other words, an effective friction damper generally must have apparent macroslip between the interfaces. Therefore, the macroslip approach has been widely accepted for modelling the friction interfaces.

In most research on the macroslip approach, the friction coefficient between the interfaces has been assumed to be constant or to have two discrete values, one representing the static friction coefficient, the other the dynamic friction coefficient. However, experimental results[6-9] have indicated that the dynamic friction coefficient is dependent on the sliding velocity of the contact surfaces, especially when stick-slip motion occurs. The fact that the friction coefficient during sliding is smaller than the static friction coefficient can be explained by imagining that the asperities on one contact surface can jump part of the way over the gap between the asperities on the other surface. As a consequence, the effective roughness is reduced. Experimental results[6-8] have shown that the friction coefficient decreases exponentially with relative velocity. A graph of a typical dry friction coefficient vs. the relative velocity of the contact surfaces is shown in Fig. 4. In this chapter, the dynamic behavior of blades with constant and variable friction coefficients will be discussed.

Two important factors may affect the damping capability of friction interfaces. The first is the normal load of the interfaces, and the other is the orientation of the interfaces relative to the direction of vibration. The orientation of the shroud interfaces of blades not only can alter the damping and stiffness coupling among the blades, but also cause the slip load to vary dynamically during the vibration cycle.

In this chapter, the modeling of the friction interface will first be discussed. The application of a friction damper to reduce the vibration of turbine blades will then be discussed in detail.

2. Dynamic Modeling of Friction Interfaces

Before one can apply a friction damper to reduce the vibration of a structure, one must be able to model the motion of the friction interfaces exactly. For simplicity, we consider a rigid body moving on a rough surface, as shown in Fig. 5. The rigid body with mass m is subjected to a normal force N and a tangential external force f. If the external force f is large enough, the rigid body will begin to move. The process from rest to motion can be separated into several steps, as shown in Fig. 6. When the tangential force f is applied, the locked asperities begin to deform and the rigid body begins micro-motion. When the tangential force becomes large enough, the asperities are completely deformed and are no longer locked together; the body then begins to slip on the surface.

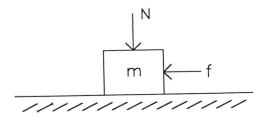

Fig. 5   A rigid body moving on friction surface

According to the model in Fig. 6, the motion of the rigid body can be represented by the model shown in Fig. 7. The mass m represents the rigid body and the block behind the mass m represents the effect of the asperities. The spring, K_d, between the block and the mass m denotes the shear stiffness of the asperities. When the motion of mass m begins, the asperities are locked together and the mass m is subjected to the spring force, $K_d(x-y)$. When a sufficiently large displacement x is reached, the spring force makes the asperities yield completely, and slipping occurs. While slipping, the mass is subjected to dry friction force, μN sign $(\dot{y})$. So the y in Fig. 7 can be regarded as the motion of the friction damper.

According to this model, the equation of motion for the rigid body can be written as

$$m\ddot{x} = f + f_n \tag{1}$$

where f is the external force indicated in Fig. 5 and f_n is the friction force due to the contact interfaces. If N and K_d represent the static normal load of the rigid body and the shear stiffness of the asperities (damper), respectively, the friction force f_n can be expressed as,

$$f_n = \begin{cases} -K_d(x-y) & \text{when } \dot{y} = 0 \\ -\mu N sign(\dot{y}) & \text{when } \dot{y} \neq 0 \end{cases} \tag{2}$$

where y is the displacement of the damper and μ is the kinetic friction coefficient. Note that when $\dot{y} = 0$, the damper (asperities) is stuck, while when $\dot{y} \neq 0$, the damper is slipping . As shown in Fig. 4, the kinetic friction coefficient is an exponential function of the relative velocity. Therefore, the kinetic friction coefficient can be expressed approximately as

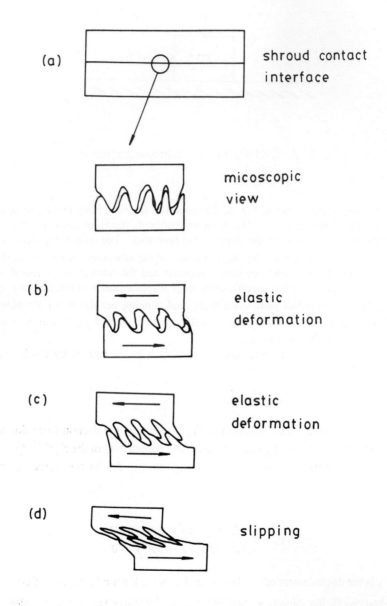

Fig. 6   Microsopic view of the friction interfaces

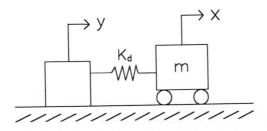

Fig. 7 Model representing the relative motion between friction inter-faces

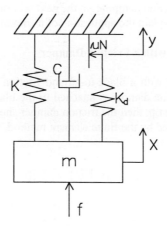

Fig. 8 Model of one-mode system

$$\mu = a + b \exp^{-d|\dot{y}|} \tag{3}$$

where a, b, and d are constants. Note that if $\dot{y} = 0$, then μ is equal to $\mu_s = a + b$. The μ_s represents the static friction coefficient. The constant a is the lower bound of the kinetic friction coefficient and $a + b = \mu_s$ is the upper bound. The constant d can be regarded as the decay rate of the kinetic friction coefficient with the sliding velocity of the damper. The actual values of a, b, and d depend on the conditions of the friction surfaces and generally can be determined only by experiments in which the friction force and the sliding velocity are measured simultaneously. However, for general metal contact surfaces, $d = 1 \sim 2$ s/m is a reasonable value.

If the damper (i.e., y) is sticking, then the velocity of the damper must be zero; if the damper is slipping, then its velocity must be equal to the velocity of the mass (i.e., x). Therefore, the term $\dot{y}$ should be

$$\dot{y} = \begin{cases} 0 & when \ |x - y| < \mu N / K_d \\ \dot{x} & when \ |x - y| \geq \mu N / K_d \end{cases} \tag{4}$$

Eqs. (2) to (4) are the equations of motion of a friction damper. They can be coupled to any structure to provide friction damping. In the next section, a single turbine blade will be used as an example to explain the application of a friction damper.

3. Single Turbine Blade with a Friction Damper

A typical single blade with a blade-to-ground friction damper is shown in Fig. 1. Fig. 1 shows that the friction damper is attached to the blade root. If the blade has an integral shroud at the blade tip, then the friction damper should be attached to the blade tip. If the blade is discretized by the finite element method, then the equation of motion of the blade can be written as

$$[M]\{\ddot{U}\} + [C]\{\dot{U}\} + [K]\{U\} = \{F\} + \{F_n\} \tag{5}$$

where {U} is the generalized displacement vector; [M], [C], and [K] are the mass, material damping, and stiffness matrices; and {F} and {F_n} are the linear external force and non-linear friction force vectors. Eq. (5) is a nonlinear equation that can be solved by direct time-step integration with iterative process, such as the Runge-Kutta method. However, Eq. (5) generally has many degrees of freedom, and thus solving this equation directly is very time consuming. Generally, a violent vibration appears only when the blade is subjected to an excitation force with a frequency near one of the

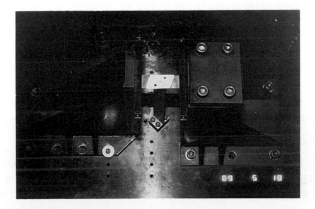

Fig. 9   Single blade with friction interfaces on both sides of the shroud

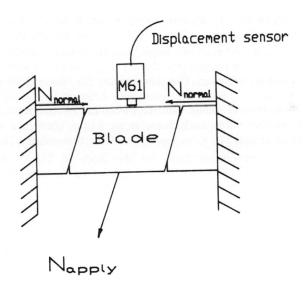

Fig. 10 Measurement of friction coefficient

natural frequencies of the blade. Therefore, equation (5) can be reduced to a single degree of freedom by one-mode approximation, as shown in Fig. 8. The equation of motion of the single-mode model can be written as

$$m\ddot{x} + c\dot{x} + kx = f + f_n \qquad (6)$$

where f is the external excitation force and f_n is the friction force. N and K_d in Fig. 8 represent the static normal preload of the friction damper and the stiffness of the damper in the direction of relative vibration. The friction force, f_n, can be expressed as Eq. (2). Note that Eq. (6) is a transformed equation; the relations between N and N_t, $\{f_n\}$ and $\{F_n\}$, μ and μ_t, and K_t and K_d can be found in Appendix A.

Although Eq. (6) is a nonlinear equation with one degree of freedom only, direct time-step integration is still very time consuming. A harmonic balance method with many harmonic terms has been proposed by Wang[10]. The main advantage of this method is its computational efficiency: it requires only about one-tenth of the computer time needed for the direct integration method.

3.1. Verification of the Friction Model

The accuracy of Eq. (5) depends mainly on the model of the friction interfaces, i.e., $\{F_n\}$. The simplest way to verify the model of the friction interfaces is to investigate the dynamic behavior of a single blade with friction interfaces. Fig. 9 shows that a single blade with an integral shroud at the tip is constrained with a known normal preload at the shroud interfaces. One can measure the responses of the blade when subjected to different exciting force levels with different normal preloads, and then compare the experimental results with the theoretical results obtained from Eq. (5). Before the comparison can be made, however, the friction coefficient and the stiffness of the friction interfaces, i.e., K_t in Fig. 1, should be determined. The static friction coefficient of the shroud interfaces for the blade in Fig. 9 was determined experimentally, as shown in Fig. 10. The interfaces of the shroud were constrained by a rigid framework with a known normal preload. A force that would cause the shroud to slide along the interfaces was applied gradually. Once a gross slip of the shroud was detected, the static friction coefficient could be determined. The static friction coefficient was found to be 0.058. The shear stiffness of the shroud interfaces was very difficult to determine experimentally. According to the experimental results of Burdekin[11], the shear stiffness of ground-ground cast iron surfaces is about 1.1×10^7N/cm. The blade in Fig. 10 was made from stainless steel, and thus the shear stiffness of the shroud interfaces can be estimated by,

$$K_t = K_{t,c} \times \frac{E_s}{E_c} \qquad (7)$$

where $K_{t,c}$ is the shear stiffness of cast iron and E_s and E_c are the elasticity moduli of stainless steel and cast iron, respectively. So the shear stiffness of the shroud interfaces was assumed to be $K_t = 1.35 \times 10^7 \, N/cm$. In the theoretical simulation, it was assumed that the friction coefficient was constant. The effect of a variable friction coefficient will be discussed later.

Figs. 11 and 12 show the forced responses as a function of excitation frequencies with normal preload $N_0 = 40N$ and excitation force amplitude $F_0 = 0.5N$ and $1N$, respectively[12]. Nonlinear behavior can clearly be observed, in that the responses are not proportional to the amplitude of the exciting forces. The results in Figs. 11 and 12 indicate that the model of the friction damper is reasonable, even though the friction coefficient is assumed to be constant. The model in Fig. 9 was used only for the verification of the friction model, it did not exist in the actual turbine. In the actual turbine, the blades are brought into contact via shrouds with neighboring blades by the oversize of the shrouds circumferentially or by centrifugal force, which tends to untwist the airfoil as the engine speed increases. Blades with integral shrouds coupled circumferentially are generally called quasi-solid blades. The problem of designing a friction damper for quasi-solid blades will be discussed in the next section.

4. Design of Friction Damper for Quasi-Solid Blades

Quasi-solid blades are coupled by shroud-interfaces. In other words, the blades are coupled together through friction dampers. So the considerations involved in designing a friction damper for these blades are different from those for a single blade. It is more difficult to control the normal preload of the friction interfaces. The values of the normal preload at the shroud interfaces may be quite different from blade to blade because of manufacturing and assembly tolerances. This problem is generally referred to as mistuning of the static preload. The responses of the coupled blades are more complicated than that of a single blade. As a consequence, it is more difficult to design friction dampers (i.e., shroud interfaces) to control the vibration modes. In this section, some of the problems involved in designing friction dampers for quasi-solid blades will be discussed.

4.1. Mathematical Formulation[13]

A typical blade with integral shroud at the tip is shown in Fig. 13. The direction from blade root to tip is defined as the Z axis, and the circumferential direction of the assembled blades is defined as the X axis. Generally, the shroud is relatively rigid in comparison with the blade itself. Hence, the shroud can be considered as a lumped

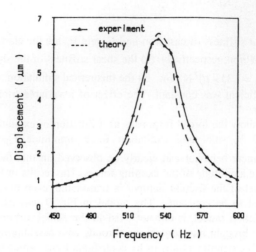

Fig. 11 Response of the blade in Fig. 9, preload $N_0=40N$, exciting force $F_0=0.5N$ (Ref. 15)

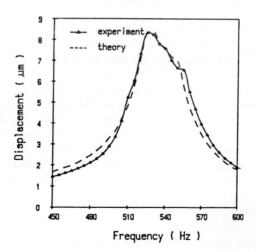

Fig. 12 Response of blade in Fig. 9, preload $N_0=40N$, exciting force $F_0=1N$ (Ref. 15)

mass with three degrees of freedom, U, V, and θ which will be coupled with the neighboring blades at the tip, as shown in Fig. 14. Here U, V, and θ represent the displacements in the X and Y directions and the rotation about the Z axis, respectively. When the blades are assembled on the disk, the intersection of X axis and the plane of the shroud interface forms a angle, β_1, as shown in Fig. 14. The displacements in the X-Y coordinate system can be transformed into a new coordinate system H-G, which is tangential and normal to the shroud interface. So the degrees of freedom of the i-th blade, U_i, V_i, θ_i, with which the blade is coupled to the neighboring blades become

$$\begin{Bmatrix} h_i \\ g_i \\ \theta_i \end{Bmatrix} = \begin{bmatrix} \cos\beta_1 & , \sin\beta_1 & ,0 \\ -\sin\beta_1 & , \cos\beta_1 & ,0 \\ 0 & ,0 & ,1 \end{bmatrix} \begin{Bmatrix} U_i \\ V_i \\ \theta_i \end{Bmatrix} \tag{8}$$

where h_i, g_i, and θ_i are the degrees of freedom in th H, G, and Z directions, respectively. The relative motion at the blade tip between the i-th blade and the (i+1)-th blade in H-G coordinates is

$$\Delta h_i = h_i - h_{i+1}$$
$$\Delta g_i = g_i - g_{i+1} \tag{9}$$

assume that the motion is restricted elastically so that the normal load at the shroud interface, N_t, is given by

$$N_{ti} = N_{oi} + K_g \Delta g_i \quad \text{for} \quad N_{ti} \geq 0 \tag{10}$$

where N_{ti} is called the dynamic normal load, and N_{ti} becomes zero when shroud separation is imminent. N_{oi} is the static compressive preload between the i and (i+1) blades, and K_g is the stiffness of the shroud interface in the G direction. According to Eq. (10), the normal load of the friction interfaces varies with the motion of the blades. This is an important characteristic of the friction damper in quasi-solid blades. The effect of the variable normal load on the dynamic behavior of the blades will be discussed later.

The elastic force at the shroud interfaces in the G direction and the friction force in the H direction vanish when shroud separation occurs, i.e., $N_{ti} \leq 0$.

If separation occurs, the blades will be subjected to impact force, which may enhance the failure of the blades. According to Eq. (10), $N_{ti} \leq 0$ only when N_{0i} is too low or the relative vibration Δg_i is too large. Thus the static normal preload N_{0i} should

Fig. 13 Blade with integral shroud at the tip

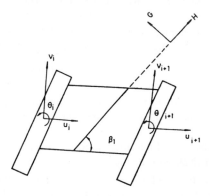

Fig. 14 Shroud geometry (top view)

be designed to prevent any separation. The friction interfaces can then be modeled as shown in Eq. (2), i.e.,

$$
f_{ni} = \begin{cases} - K_d (h_i - y_i) & when \left| h_i - y_i \right| < \mu N_{ti} / K_d \\ -\mu N_{ti} \, sign(\Delta \, \dot{h}_i) & when \left| h_i - y_i \right| \geq \mu N_{ti} / K_d \end{cases} \tag{11}
$$

where K_d represents the shear stiffness of the interfaces in the H direction. For simplicity, assume that the friction coefficient μ is constant. Then $\dot{y}_i$ should be

$$
\dot{y}_i = \begin{cases} \dot{h}_{i+1} & when \left| h_i - y_i \right| < \mu N_{ti} / K_d \\ \dot{h}_i & when \left| h_i - y_i \right| \geq \mu N_{ti} / K_d \end{cases} \tag{12}
$$

With the model of the shroud interfaces, the system equation of motion of the quasi-solid blades can be written as

$$
[M]\{\ddot{U}\} + [C]\{\dot{U}\} + [K]\{U\} = \{F\} + \{F_n\} \tag{13}
$$

where $\{F_n\}$ is the nonlinear friction force vector. Solving Eq. (13) directly by numerical iteration is impractical, because the nonlinearity arises only from some local points in the structure, i.e., the shroud interfaces. An effective computational procedure based on the concept of component mode synthesis[14] can be adopted to reduce the DOF of Eq. (13). First, one divides the assembled blades into N identical substructures, i.e., each single blade represents a linear substructure. The interior displacements of each substructure can be expressed in terms of the boundary displacements (i.e., the displacement of the shroud) by static condensation and supplemented by a few lowest modes of the substructure. The condensed equations of motion of all the substructures are then coupled together through the nonlinear joints (the shroud interfaces).

The vector $\{F\}$ in Eq. (13) is the external force. The most important external force for the turbine blades is the nozzle wake flow. The nozzle wake excitation can be expressed in the form of Fourier series as

$$
f_{t,i} = \sum_{k-1}^{\infty} A_k \cos(k\Omega t + \alpha_{k,i}) \tag{14}
$$

$$
\alpha_{k,i} = \frac{2 \pi k (i - 1)}{N_b}
$$

where $f_{t,i}$ is the nozzle wake excitation force acting on the i-th blade in the tangential direction (X-direction), Ω is the rotational speed of the disk, and N_b is the total number of blades on the disk.   Similarly, the axial force and moment in the Z direction may have the same form as that in Eq. (14).   The forced vibration of quasi-solid blades will be discussed below.   First, however, some design considerations relevant to friction dampers are discussed briefly in the next section.

4.2. Design Parameters of a Friction Damper

There are two basic parameters that control the characteristics of a friction damper, i.e., μ and N.   The friction coefficient μ depends mainly on the material and roughness of the contact interfaces.   It is difficult to design a desired value of μ to obtain an optimal friction damper.   Hence the normal load N is the most important parameter in designing a friction damper.

The normal forces of quasi-solid blades may vary from blade to blade, as indicated by Eq. (10).   However, results reported in the literature[13] indicate that if lift-off (separation) does not occur between the interfaces, the responses of quasi-solid blades predicted by using variable normal load are nearly the same as those predicted by constant normal load.   Because lift-off may cause serious impact excitation, the static preload N_0 generally is designed to prevent possible lift-off.   In other words, for simplicity the model of constant normal load can be used in practice.   If the model of constant normal load is used, then Eq. (13) can be normalized with respect to external excitation force.   In other words, Eq. (13) can be written in dimensionless form in a manner such that the slip load is divided by the excitation force.   The vibration amplitude will then vary according to the static normal load N_0 for a constant excitation force or according to the excitation force for a constant static preload.   So, the normalized slip load $\mu N_0 / F_0$ is the first important design variable, where N_0 is the static normal load and F_0 is the amplitude of the external excitation force. Note that each blade of the quasi-solid blade system is subjected to an external force with same amplitude but with different phase angle, as indicated in Eq. (14).   This is why a single value of $\mu N_0 / F_0$ can be found for quasi-solid blades.   If the external excitation frequency is known, it is possible to find an optimal value of $\mu N_0 / F_0$ at which the blade system has minimum vibration amplitude.   However, a friction force is provided by the shroud interface, which has a certain orientation relative to the tangential direction (X-axis), as shown in Fig. 14.   Different orientations of the interface will provide different friction forces for different directions.   Therefore, the optimal value of $\mu N_0 / F_0$ can't be determined without considering the orientation of the interface.

In summary, the normalized slip load $\mu N_0 / F_0$ and the orientation of the shroud interface are the two most important design factors of a friction damper for quasi-solid blades. The design of these two factors will be discussed in detail in the next section.

4.3. Design of Optimal Interface Angle[15]

The shroud interface has two basic functions: to provide friction damping and to provide stiffness coupling among the blades. Thus, the angle of the shroud interface can affect both the damping and coupling stiffness. This is why the interface angle is an important design factor in a friction damper.

The shroud interface generally is a plane surface for manufacture simplicity. An angle, β, must first be defined to indicate the orientation of the shroud interface. The angle β is defined as the angle formed by the shroud interface and the direction of vibration of the shroud of a free-standing single blade at its first natural frequency. For example, if the vibration direction of the shroud of a single blade at its first natural frequency is -12 degrees relative to the X axis (the tangential axis), then the angle β is

$$\beta = \beta_1 - (-12^0)$$

where β_1 is as defined in Fig. 14.

Before discussing the optimal design of the angle β, we must explain two important characteristics of the forced response of quasi-solid blades excited by the wake flow. The first characteristic is that all blades experience the same vibration amplitude at a given excitation frequency. The second characteristic is that the vibration mode with all blades vibrating in phase cannot be excited except when k/N_b in Eq. (14) is an integer. Thus if vibration due to wake excitation is a serious concern, one should pay more attention to the out-of-phase resonant vibration. A pretwisted blade generally has a bending-bending or bending-torsion coupled vibration mode. However, if the bending vibration is dominant in the tangential direction, then for simplicity the vibration mode can be considered as a tangential mode. In the following discussion, the tangential out-of-phase vibration and axial out-of-phase vibration will be discussed separately. The blade used for simulation has a pretwisted and tapered cross-section.

The first out-of-phase tangential vibration mode will be used here to show how to minimize the resonant vibration by properly designing the angle of the shroud interface. The first out-of-phase tangential mode has a maximum deflection at the midpoints of the blades, and the shrouds remain nearly stationary without displacement. It is useful to normalize the maximum vibration of the quasi-solid blades, Λ_q, by the maximum amplitude of the first tangential mode, A_0, of a free-standing single blade subjected to the same excitation force. Because the free-standing blade does not have coupling, the normalized value can show the effect of coupling. The normalized maximum

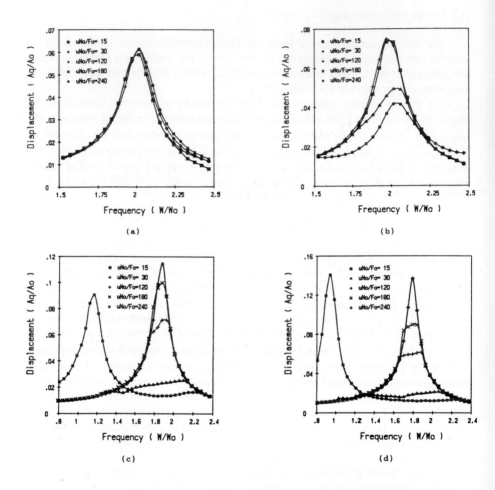

Fig. 15 Response spectra for different slip loads, (a)$\beta=90°$; (b)$\beta=70°$
; (c)$\beta=40°$; (d)$\beta=10°$ (Ref. 15)

amplitude A_q/A_0 is shown in Fig. 15 for β=10, 40, 70 and 90 degrees as functions of the normalized excitation frequency, ω/ω_o, with normalized slip load $\mu N_0/F_0$ as parameter. Here, A_0 is the blade tip displacement of the first tangential mode of a free-standing single blade subjected to the same excitation force as the quasi-solid blades, and ω_o is the first natural frequency of the free-standing single blade.

The result in Fig. 15(c) shows that the peak responses near ω/ω_o=1.9 are gradually reduced with decreasing static preload until no apparent peak appears. If the static preload is further decreased, a new peak response will appear near ω/ω_o=1.1. The appearance of a new peak near ω_o, which will be referred to here as the "weak coupling effect", occurs when the normalized slip load (or static preload) is below a certain value. The characteristics of the "weak coupling effect" are: (1) each blade behaves like a free-standing single blade; (2) a large slip displacement occurs at the shroud interfaces; (3) the peak frequency is near the first natural frequency of the free-standing single blade. How to select a minimum static preload so as to avoid the weak coupling effect will be discussed later; first we will discuss the effect of the angle β on the resonant responses. If the static preload is high, then the resonant response is smaller for a larger angle β. Conversely, if the static preload is low, then the resonant response is smaller for a smaller angle β. To clarify the effect of β on the resonant responses, the maximum responses were calculated as a function of the normalized friction force, $\mu N_0/F_0$, for different value of β. The results are shown in Fig. 16.

One can see that β=10^0 has the smallest amplitude in the range from $\mu N_0/F_0$=30 to 120, which will be referred to below as the "damping region". The reason for this result is that a small β will provide larger damping in this range, because the direction of vibration is more nearly parallel to the shroud interface for a small angle β. However, if the preload is increased to a higher range, the stiffness region, then no slip will occur and the vibration will be controlled by the stiffness in the tangential direction. A small angle β angle has a soft stiffness in the tangential direction provided by the shear stiffness of the shroud interfaces. As a consequence, a small β may cause a large response. Note that the shear stiffness K_d generally is smaller than K_g. The above discussion leads to the following conclusion: to minimize the out-of-phase tangential resonant vibration, a small angle β and a small static preload should be selected; however, the static preload should not be so small as to cause weak coupling to occur. The remaining question is how to determine the smallest static preload, called the critical preload, so as to avoid the weak coupling effect.

Because the weak coupling effect is accompanied by slip between the shroud interfaces, the critical preload, N_{cr}, is the smallest static preload that will prevent the slipping under a certain excitation force. So N_{cr} must satisfy the following equation:

$$\mu N_{cr} = K_d \, \Delta h_m \tag{15}$$

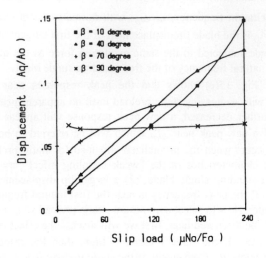

Fig. 16 Maximum responses of the out of phase tangential resonant
vibration versus slip loads for different β angles

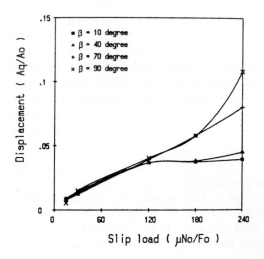

Fig. 17 Maximum responses of the out of phase axial resonant vibration versus slip loads for different β angles

where Δh_m is the maximum elastic displacement of the shroud (caused by the elastic deformation of the asperities before slipping) between the shroud interfaces in the H direction. According to the model of Eq. (11), if there is no slipping between the shroud interfaces, the structure of the quasi-solid blades is linear. If the number of blades N_b, the speed order k in Eq. (14), and K_d are determined, then Δh_m is proportional to the magnitude of the excitation force in the H direction, F_β, because of the linearity of the structure. However, if N_b and k are variables, and K_d and F_β are fixed, then Δh_m is proportional to N_b/k. Furthermore, if K_d is a variable, then Δh_m is proportional to $1/K_d$, because the displacement is proportional to the reciprocal of the stiffness. Therefore, if F_β, N_b, k, and K_d are all variables, then Δh_m is proportional to $F_\beta N_b/kK_d$. From Eq. (15) one has,

$$
\begin{aligned}
N_{cr} &= \frac{K_d \, \Delta h_m}{\mu} \\
&\propto \frac{K_d}{\mu} \cdot \frac{F_\beta N_b}{k \, K_d} \\
&\propto \frac{F_\beta N_b}{\mu k}
\end{aligned}
\tag{16}
$$

where $\propto$ represents the proportional relationship. Note that K_d bears no relation to the critical preload. The result indicates that if the number of blades is large or the harmonic order of the wake excitation is small, the weak coupling effect may easily be induced by insufficient preload. We can conclude that: to minimize the out-of-phase tangential resonant vibration, a small angle β and a static preload somewhat larger than N_{cr} should be selected.

The behavior of the out-of-phase axial resonant vibration is similar to that of the out-of-phase tangential resonant vibration. Fig. 17 shows the maximum response of the out-of-phase axial resonant vibration as a function of the normalized friction force for different values of β. One can see that when the friction force is large, a larger β will cause a larger response. This is because in the stiffness region K_d offers softer stiffness coupling in the axial direction for a large angle β. Note that K_d represents the shear stiffness of the asperities and generally is smaller than K_g. The result in Fig. 17 also shows that a small static preload will reduce the response for all values of β. This is quite different from the case of the tangential resonant vibration. The reason for this result is that the stiffness coupling in the H direction is always smaller than that in the G direction, so energy can be dissipated due to significant relative displacement of the interfaces for all values of β in the damping region. Of course, a larger β will make the damping of the interfaces more efficient, because when β is large the axial vibration is more nearly parallel to the shroud interfaces. However, a small β will offer stiffer coupling in the axial direction, which will reduce the maximum response.

4.4. Conclusions

The quasi-solid blades are brought into contact via the integral shrouds by the oversize of the shrouds circumferentially or by centrifugal force, which tends to untwist the airfoil as the turbine speed increases. The damping effect provided by the shroud interfaces depends on three basic parameters, i.e., μ, N_0, and the orientation of the interfaces. The angle (orientation) of the shroud interface can alter not only the damping but also the coupling between the blades. Hence the angle of the shroud interface is a very important design factor for controlling the vibration of the blades. Generally, a small interface angle is superior than a large interface angle in minimizing the resonant vibration caused by wake flow. However, a critical static preload proportional to $F_\beta N_b / \mu$ k should be selected to prevent weak coupling effect. A small shroud angle has the above mentioned advantage, but it may be more sensitive to preload mistuning than a large shroud angle[16]. To avoid this shortcoming, a more complicated shroud interface can be used, such as that shown schematically in Fig. 18, which is no longer a simple plane. Of course, the manufacturing cost of a more complicated shroud interface may be higher, and the contact normal force may become more difficult to control.

5. Effect of Variable Friction Coefficient[17]

Eq. (3) shows that the friction coefficient is a function of the relative velocity of the contact surface. If the friction coefficient varies with relative velocity, then Eq. (6) becomes more difficult to solve than when μ=constant. The harmonic balanced method[10] can no longer be applied, and generally a high-order Runge-Kutta method or an improved Newton-Raphson method should be used. Of course, both of these methods are very time-consuming. In this section, the effect of a variable friction coefficient on the dynamic behavior of blades with friction dampers will be discussed briefly.

Generally, there are two issues which are of most interest to a damper designer: the optimal normal load of the damper and the resonant frequency shift due to different normal loads. As discussed before, the optimal normal load is the normal load with which the blade has minimum resonant reponse. The frequency shift indicates the difference between the resonant frequencies of the free-standing blade (N=0) and the

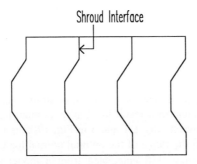

Fig. 18 Shroud with complex contact surface

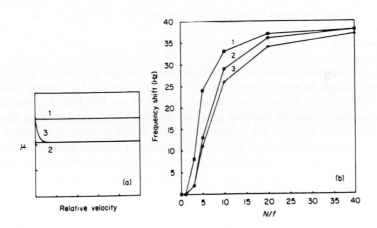

Fig. 19 (a) Three different models of friction coefficient vs. relative velocity; (b) the frequency shift calculated with three different friction coefficients 1. $\mu = 0.30$; 2. $\mu = 0.21$; 3. $\mu = 0.21 + 0.09$ $\exp(-0.15|\dot{y}|)$

frictional damped blade.   In the following discussion a single blade with a friction damper will be considered.   If a single-mode model is used, then Eq. (6) is the equation of motion. Fig. 19(b) shows the frequency shift due to different normal loads for the three different friction coefficients shown in Fig. 19(a). Note that the normal load N is normalized by the external excitation force f. Although the average value of the variable friction coefficient (curve 3 in Fig. 19(a)) is between the values of the two constant coefficients (curves 1 and 2 in Fig. 19(a)), the frequency shift calculated with the variable friction coefficient (curve 3 in Fig. 19(b)) is smaller than that calculated from the constant friction coefficients (curves 1 and 2 in Fig. 19(b)). This result indicates that if the actual friction coefficient is a function of sliding velocity, then a model with a constant friction coefficient may result in a faulty resonant frequency. In addition, the maximum difference between curves 1 and 3 in Fig. 19(b) is found in the range N/f=5 $\sim$ 10, which falls right in the range of the optimal normal load. The result in Fig. 19(b) indicates that the frequency shift calculated from a model with a variable friction coefficient is always smaller than the frequency shift calculated from a model with a constant friction coefficient.   In addition, the resonant response predicted by a variable friction coefficient model is higher than that predicted by a constant friction coefficient model.

Although it is more difficult to measure the friction coefficient as a function of sliding velocity, an accurate friction coefficient is necessary for designing a correct friction damper.   However, it is difficult to find data on friction coefficients in the literature. A further research work should be promoted in this area.

6. Conclusion

A real machine generally consists of many components which are connected together through different joints. Most of the joints can provide a certain amount of friction damping to the machine. Indeed, one might say that friction damping is the most important source of damping in a real machine. However, the main function of most mechanical joints is not to provide friction damping. A friction damper can be regarded as a special kind of joint.   Theoretically, a friction damper is a powerful device for controlling the vibration of a structure. In practice, only a few examples of friction dampers have been used successfully. Most of the friction damping is provided by dry friction surfaces[18].   It is very difficult to design a friction joint which has apparent relative motion, however, without adversely affecting the behavior of a real machine. This is why it is not easy to design a friction damper to control the vibration of a machine.   The friction dampers used with turbine blades are one of the few successful practical examples of such dampers.   The friction model and design considerations for a friction damper introduced in this chapter can be applied to general problems concerning friction interfaces. The one-mode model in Fig. 8 generally can be applied to study the vibration of any structure with friction interfaces. The transformation from a complex structure to the one-mode model can be found in Appendix A. The most important design factors of a friction damper are the friction coefficient μ, the normal load N, the shear stiffness of the friction asperities, the orientation of the friction interfaces, and the external excitation force f. The characteristics of these factors are summarized as follows:

(1) The friction coefficient μ generally is an exponential function of the relative velocity of the friction interfaces. The main problem is that it is very difficult to know the exact function between μ and the relative velocity. For a preliminary design, the friction coefficient can be assumed to be constant and equal to the static friction coefficient.

(2) The normal load N may vary with the operating conditions of the machine. For instance, quasi-solid blades are brought into contact via an integral shroud by the oversize of the shrouds circumferentially. However, the normal contact force (prestress) may change due to circumferential expansion caused by centrifugal forces or temperature. Hence the effect of operating conditions on the normal load should be considered carefully in advance.

(3) The shear stiffness of the friction interfaces is on about the order of 1.0×10^7 N/cm for metal surfaces.

(4) The orientation of friction interfaces can affect the coupling strength between the interfaces and the damping capability of the interfaces. The orientation of the friction interfaces should be determined according to the direction of the relative vibration of the interfaces.

7. References

1. C. M. Menq, J. Bielak, and J. H. Griffin, *Journal of Sound and Vibration*, **Vol. 107** (1986), p. 279.

2. C. M. Menq, J. H. Griffin, and J. Bielak, *ASME, Journal of Vibration, Acoustics, Stress, and Reliability in Design*, **Vol. 108** (1986) p. 50.

3. A. Muszynska and D. I. G. Jones, *Shock and Vibration Bullitin*, (1979) p. 89.

4. H. B. Schwarz and E. H. Dowell, *Journal of Sound and Vibration* , **Vol. 91** (1983) p. 269

5. J. H. Griffin, ASME, *Journal of Engineering for Power*, **Vol. 102** (1980) p. 329

6. J. B. Sampson, F. Morgan, D. W. Reed and M. Muskat, *Journal of Applied Physics*, **Vol. 14** (1943) p. 689.

7. J. T. Oden and J. A. C. Martins, *Computer Methods in Applied Mechanics and Engineering*, **Vol. 52** (1985) p. 529.

8. K. Kraft, *AET 22 addendum to ETR*, (1967) p. 58.

9. E. Schneider and K. Popp, *Journal of Sound and Vibration*, **Vol. 120** (1988) p. 227.

10. J. H. Wang and W. K. Chen, *ASME, Journal of Engineering for Gas Turbines and Power*, **Vol. 115** (1993) p. 294.

11. M. Burdekin, N. Back, and A. Cowley, *Journal of Mechanical Engineering Science*, **Vol. 20** (1978) p. 129.

12.   J. H. Wang and H. L. Yau, *6 th   National Conference on Mechanical Engineering*, Taiwan (1989) p. 997.
13.   H. L. Yau, *Master's Thesis, National Tsing Hua University*, Taiwan (1989).
14.   R. W. Clough and E. L. *Wilson, Computer Method in Applied Mechanics and Engineering*, **Vol. 17/18** (1979) p. 107.
15.   J. H. Wang and H. L. Yau, *Gas Turbine and Aeroengine Congress and Exposition, Brussels, Belgium* (1990) paper No. 90-GT-247.
16.   J. H. Wang, *3rd International Conference on Rotordynamics, Lyon, France* (1989).
17.   J. H. Wang and W. L. Shieh, *Journal of Sound and Vibration*, **Vol. 149** (1991) p. 137.
18.   K. Popp, Z. angew. *Math. Mech.*, **Vol. 74** (1994) p. 147.

Appendix A

A blade with a typical friction damper is shown in Fig. 1. If the blade is discretized with the finite element method, then the equation of motion can be written as

$$[M]\{\ddot{U}\} + [C]\{\dot{U}\} + [K]\{U\} = \{F\} + \{F_n\} \tag{A-1}$$

Let $\{\Phi_i\}$ be the i-th eigenvector of Eq. (A-1) without considering the damping, i.e.,

$$[K]\{\Phi_i\} = [M]\omega_i^2\{\Phi_i\} \tag{A-2}$$

where ω_i represents the i-th eigenfrequency. The displacement vector $\{U\}$ can then be represented in terms of the eigenvectors as

$$\{U\} = \sum_{i=1}^{n} x_i(t)\{\Phi_i\} \tag{A-3}$$

If only one mode approximation is considered, say the first mode, then one can substitute Eq. (A-3) into Eq. (A-1) and then multiply by $\{\Phi_1\}^T$ to obtain

$$\{\Phi_1\}^T[M]\{\Phi_1\}\ddot{x}_1(t) + \{\Phi_1\}^T[C]\{\Phi_1\}\dot{x}_1(t) + \{\Phi_1\}^T[K]\{\Phi_1\}x_1(t)$$
$$= \{\Phi_1\}^T(\{F\} + \{F_n\}) \tag{A-4}$$

or

$$m\ddot{x}_1(t) + c\dot{x}_1(t) + kx(t) = f + f_n \qquad \text{(A-5)}$$

where

$$m = \{\Phi_1\}^T [M]\{\Phi_1\}$$

$$c = \{\Phi_1\}^T [C]\{\Phi_1\}$$

$$k = \{\Phi_1\}^T [K]\{\Phi_1\}$$

$$f = \{\Phi_1\}^T \{F\}$$

$$f_n = \{\Phi_1\}^T \{F_n\}$$

As shown in Fig. 1, the friction damper is connected to the blade at only one point. Let the connecting point be the p-th degree of freedom, the corresponding component of the $\{U\}$ vector be u_p, and the corresponding component of $\{\Phi_1\}$ vector be ϕ_{1p}. Then f_n can be written as

$$f_n = \begin{cases} -\phi_{1p} \cdot K_t \cdot (u_p - y_d) & \text{when } \dot{y}_d = 0 \\ -\phi_{1p} \cdot \mu_t N_t sign(\dot{u}_p) & \text{when } \dot{y}_d > 0 \end{cases} \qquad \text{(A-6)}$$

where μ_t, N_t, y_d, K_t are shown in Fig. 1. However, according to Eq. (A-3), one has

$$u_p = \phi_{1p} x_1(t) \qquad \text{(A-7)}$$

So Eq.(A-6) can be rewritten as

$$f_n = \begin{cases} -K_d(x_1 - y) & \text{when } \dot{y} = 0 \\ -\mu N sign(\dot{y}) & \text{when } \dot{y} > 0 \end{cases} \qquad \text{(A-8)}$$

where

$$K_d = \phi_{1p}^2 \cdot K_t$$

$$y = y_d / \phi_{1p}$$

$$\mu N = \phi_{1p} \mu_t N_t$$

Eq. (A-8) is the same as Eq. (2), and the transformed parameters K_d, y, μN are shown in Fig. 8.

Dynamics with Friction: Modeling, Analysis and Experiment, pp. 197–232
edited by A. Guran, F. Pfeiffer and K. Popp
Series on Stability, Vibration and Control of Systems Series B: Vol. 7
© World Scientific Publishing Company

MODELING UNSTEADY LUBRICATED FRICTION

ANDREAS POLYCARPOU AND ANDRES SOOM
Department of Mechanical and Aerospace Engineering
State University of New York at Buffalo, 321 Gregory B. Jarvis Hall
Buffalo, NY 14260, USA

ABSTRACT

Friction models that relate normal and friction forces at sliding contacts are widely
used in the design and analysis of many mechanical systems and components. Most
models include many hidden assumptions and tend to be valid under rather limited sets
of conditions. Studies of unsteady sliding, (i.e., when some combination of sliding
speed, applied load, friction force or friction coefficient vary with time), can reveal
some of the limitations of simplified models. In this paper we summarize a series of
studies of unsteady sliding at a liquid lubricated line contact. The unsteadiness arises
due to changes in sliding velocity or normal load, and sometimes includes stick-slip.
The lubrication takes place in the mixed or boundary regimes, so that the frictional
resistance is developed due to a combination of fluid shear over the entire contact and
solid friction at localized asperity contacts. A number of parameters are varied and
the key elements of any dynamic friction model are identified. It is suggested that a
geometric approach to the problem, wherein the instantaneous friction is related to the
instantaneous separation of the sliding bodies and the sliding velocity, can capture
most aspects of friction behavior.

1. Introduction

Many important mechanical devices and machines include sliding contacts or interfaces
which must operate under transient, oscillatory or other dynamic conditions. The design and
analysis of such dynamic systems has become increasingly sophisticated. For example,
control systems may require greater bandwidth or finer precision. When addressing noise
problems such as squeaks and chirps, high frequency contact vibrations must be analyzed.
Historically, contacts have usually been assumed to be rigid and all relevant contact behavior
describable by a friction coefficient. Such models, will not always capture realistic contact
behavior and more sophisticated approaches, that can account for the stiffness and damping

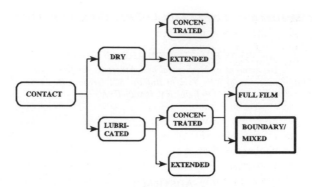

Fig. 1   Types of interface contacts.

characteristics of the interface, along with more accurate friction relations, are likely to be needed.   While friction coefficients may adequately describe the average or steady-state behavior of sliding contacts under a particular set of operating conditions, they often fail to capture transient or unsteady friction behavior over a wide range of conditions.

In discussing and developing such models of frictional and contact phenomena, one should make a number of distinctions, as illustrated in Fig. 1.   Different approaches to friction modeling are appropriate to the various categories listed. The first distinction is between dry and lubricated sliding.   As we know from elementary mechanics and generations of design experience, dry sliding can often be described adequately by a friction coefficient, although its value may not be known in advance. However, we also know that the mechanical properties of the interface region will be significantly different from the bulk, due primarily to the fact that so-called engineering surfaces are not smooth.   An important, if not universal, concept in the mechanics of friction, articulated by Bowden and Tabor[1], is the idea that the friction force is proportional to the real area of contact.   It forms the basis for the friction coefficient being constant for extended rough contacts, since under certain assumptions, the real area of contact can be shown to be proportional to the normal load[2]. This proportionality will not always hold.   For example, when two smooth curved elastic bodies are pressed together, forming a concentrated contact, the area of contact will usually increase more slowly than the normal load, and the friction coefficient will decrease with load.   Except at very light loads, dry concentrated contacts behave more like smooth surfaces, even though they may be somewhat rough.   This assumption has been applied, with some success to describe both steady and unsteady (instantaneous) friction behavior[3,4,5].

Also, contact regions possess tangential stiffness and damping.   A constitutive local friction law, relating tractions rather than forces, that combines nonlinear normal and tangential stiffness and damping with a frictional boundary condition, has been proposed by Oden and

Martins[6,7]. The parameters of this law can be selected in such a way that it reduces to a Coulomb relation, i.e., shear traction proportional to normal stress, under steady sliding conditions. The contact stiffness and damping are modeled as power laws. A disadvantage of the Oden and Martins approach is that one may need as many as eight parameters, (instead of just a single friction coefficient), to characterize a contact. Yet some of the more demanding contact dynamics problems may demand such an approach. It is becoming more widely recognized that contact compliance plays an important role in high frequency problems such as brake squeal. In this paper we will only address solid friction to the extent that it represents one component of the total friction.

The characterization of lubricated friction tends to be more complicated, yet technologically more important than dry friction. When a full fluid film is present and there is no solid to solid contact, hydrodynamic or elastohydrodynamic lubrication theory, (e.g., Hamrock[8]) can be used to predict the load-carrying capacity and friction for most engineering applications, although time dependent problems represent only a small fraction of the tribology literature.

However, when sliding speeds or viscosities are low, or loading is very high, only partial fluid films may exist. These are the regimes of mixed or boundary lubrication. The load is carried by a combination of fluid film and solid asperity contacts. Conceptually, the total friction force or torque, arises as the sum of the solid and fluid friction. The details of surface or asperity interactions become important, and general solutions, that employ continuum approaches, are not usually available. In any case, many design applications must be completed without such detailed analyses and friction models are often used. This is typically a single sliding friction coefficient for dry sliding or a friction coefficient that depends (only) on sliding velocity for lubricated contacts. Before proceeding, it is instructive to review some previous approaches to lubricated friction modeling.

2.   Background

Most past friction modeling efforts to characterize the boundary and mixed regimes are quasi-steady, single-valued friction-velocity relations. Such depictions of friction-velocity relations are often referred to as Stribeck curves[9]. Stribeck suggested the parameter $\eta_i V/W$, where η_i, V, and W are the lubricant dynamic viscosity at the inlet conditions, sliding speed and normal load, respectively, to characterize the various lubrication regimes. However, this parameter, known as the Stribeck parameter, does not collapse the friction data except in the hydrodynamic regime. McKee and McKee[10], suggested a somewhat different normalization that collapses quasi-steady friction data somewhat better. Their dimensionless parameter, called the modified Stribeck parameter, is $\eta_i V/(WE)^{0.5}$, where E is the Young's modulus of elasticity. The $(WE)^{0.5}$ term can be viewed as proportional to the average contact pressure that would be applied to a Hertzian line contact under the load W. Neither the Stribeck nor the modified Stribeck parameters explicitly include the normal separation of the sliding bodies.

Schipper et al.[11] and Schipper and Gee[12], suggested a different dimensionless parameter called the lubrication number, $L = \eta V/(pRa_t) = H/Ra_t$, (where p is the mean Hertzian contact pressure, Ra, the combined CLA surface roughness and H an operational parameter), in their recent work on concentrated lubricated contacts. The operational parameter, H, can be related to the theoretical film thickness for smooth surfaces. That is, the lubrication number is closely related to the lambda ratio, $\Lambda = h_o/\sigma$, where h_o is the theoretical film thickness and σ the combined RMS surface roughness. In their work, which is applicable to steady-state conditions, they suggested a friction coefficient model for the mixed lubrication regime. This model is a function of the lubrication number and empirical parameters determined from a generic friction coefficient versus lubrication number curve. The parameters of the model depend upon the transition points between different lubrication regimes, and therefore are specific to particular conditions, e.g., load, geometry and viscosity, limiting their generality, even for a given contact.

Johnson et al.[13] combined the Greenwood and Williamson[14] theory of contact of random rough surfaces, with elastohydrodynamic lubrication (EHL) theory to account for the load sharing between hydrodynamic pressure and asperity contact. They did not, however, calculate the friction forces.

A full dynamic model of rough sliding contacts, operating in the mixed lubrication regime, was suggested by Rohde[15] in his work on piston ring lubrication. The friction is due to viscous lubricant shear and the shearing of surface films at the asperities. The viscous shear is calculated from hydrodynamic theory using the Patir and Cheng[16] model, which is based upon an averaged Reynolds equation. For the asperity interactions the Greenwood and Tripp[17] theory was used. His detailed numerical analysis requires micro-EHL parameters that are not readily available and also is specific to piston ring lubrication. Nevertheless, this is the first major work that included all the key variables necessary to model dynamic mixed friction.

Hess and Soom[18] modeled dynamic friction behavior under unsteady continuous sliding conditions, using a constant time delay to capture the lag between a change in sliding velocity and the corresponding change in friction. However, the time delay model does not apply to general dynamic changes in velocity or load, but only to the particular conditions under which it was measured. Furthermore, the time delay model does not explicitly include normal motions which are the principal cause of the deviation from steady-state sliding.

Harnoy et al.[19] developed a friction model with separate solid friction and fluid shear components, but for a different tribological situation. Their friction model, which is for a short journal bearing, has a fluid component developed from hydrodynamic theory and an empirical solid friction component. They simulated a number of unsteady sliding conditions and produced results which, in a number of cases, qualitatively agree with the measurements of Hess and Soom[18].

Extensive work on EHL traction by Johnson and coworkers[20-23] and Bair and Winer[24] resulted in some constitutive relations for traction. These rheological models relate the shear stress developed in the lubricant film to the shear strain rates, which are imposed on elements of the fluid by the action of sliding or rolling.

Evans and Johnson[21-23], extracted values of the fluid properties from experimental data. Using the extracted properties of the fluids, and the imposed conditions of load, speed and temperature at the line contact, "maps" were constructed with four different regimes: (I)

Newtonian, (II) Eyring, (III) Viscoelastic and (IV) Elastic-plastic.   For each regime constitutive equations are proposed and used to predict traction.

In this paper, we will outline the factors that should be considered in the detailed modeling friction at concentrated fluid lubricated contacts operating under mixed and boundary lubrication under unsteady sliding or loading.   In mixed lubrication, the load is carried primarily by the fluid with solid to solid interaction also contributing significantly to the friction.   In boundary lubrication, the load is carried primarily by the solid asperities with the fluid shear still contributing significantly to the friction. Traditionally, the fluid shear has not been thought to be important in the boundary lubrication range, but we have found it can seldom be ignored if a lubricant is present within the contact.   These sliding and lubrication conditions therefore include all the complexities of both solid and fluid friction.   We show how simple friction-velocity models break down and how a qualitative understanding of the mechanics of the problem can assist in formulating more accurate models.

The fluid friction force arises via the shear of the fluid film within a contact so that:

$$F_f = \tau_f A_f = \eta \frac{dV}{dh_o} A_f \tag{1}$$

F_f is the fluid friction force; τ_f is the fluid shear stress; A_f the fluid area of the contact; and h_o the film thickness.   The shear stress can be estimated as the product of the viscosity and the shear rate, dV/dh_o.   At this level we consider a spatially averaged viscosity, η, and a shear rate that is the sliding velocity divided by the average film thickness.   These are not, however, independent parameters.   The viscosity depends upon temperature, and, to a lesser extent, pressure.   The area is also not unambiguously defined, but can be thought of as a total or apparent contact area.   Film thickness depends on the load, speed and viscosity as well as their time histories, in particular since squeeze film effects may also be occurring.   When elastic deformations of the contact become important, as during certain elastohydrodynamic lubrication regimes, the area of fluid being sheared will depend on the load, the curvature of the surfaces and their elastic properties.   One can therefore expect that the fluid friction can be expressed as

$$F_f = F_f(V, W, \eta_i, \alpha, h_o, E', r', T, b) \tag{2}$$

where α is the pressure-viscosity coefficient; r' is the combined radius of curvature; E' the effective elastic modulus; T the temperature at the contact and b the length of the contact.

The solid component of the friction can be thought to arise due to the shear or rupture of the real area of solid contact, so that:

$$F_s = \tau_s A_r \tag{3}$$

F_s is the solid friction force; τ_s the shear strength; and A_r the real area of contact.   The shear strength of the interface material is the relevant a material property, but probably somewhat different from the bulk.   The real area of contact is formed at a number of locations, a few or hundreds, where asperities come into contact.   The shear strength is not known precisely and neither is the contact area, which depends on the load, the separation of the surfaces, the surface texture, the elastic properties of the components and their curvatures.   Therefore, the solid friction force can be written, in functional form as

$$F_s = F_s(W, h_o, E', r', \sigma, b) \qquad (4)$$

To model the total friction we add the two forces and divide by the total normal load in order to retain a friction coefficient, μ, form of relation. Therefore,

$$\mu = \frac{F_f + F_s}{W} = \mu(V, W, \eta_i, h_o, \alpha, E', r', T, \sigma, \tau_s, b) \qquad (5)$$

From the above, it should be clear that one should not expect a simple friction coefficient model that depends only on sliding speed to work under any but the most narrowly defined conditions. All of the cited variables or parameters must appear explicitly or implicitly (through coefficients) in a friction model. In the studies that we have done on a line contact we have taken the approach of significantly perturbing the load or sliding speed about some average load or sliding speed with a given lubricant and seeking functional relationships that can capture the friction behavior under the imposed time-varying conditions. A number of average loads, sliding velocities and lubricants were tested. The models that are presented cover a range of operating conditions and lubricants. The geometry, including the surface texture, the ambient temperature and the materials of the sliding bodies were not changed. We find that the separation becomes the key parameter that, when known, can unify both equilibrium and non-equilibrium friction behavior. This is due to the fact that the separation of the surfaces directly affects the shear rate, important for the fluid friction, and has a one to one functional relationship with the real area of solid contact, which is proportional to the solid friction.

3.   Experimental Procedure

3.1.   Test Apparatus

The test apparatus used to perform the experiments in this work, is described by Polycarpou and Soom[25,26]. It consists of two subsystems, a rotational system and a nominally stationary rider assembly. The rider assembly holds a cylindrical sample that forms a line contact with the edge of a rotating disk. The contact geometry is illustrated in Fig. 2. The thickness of the disk is 25.4 mm. The diameter of the insert is 19 mm and its thickness is 6 mm. The length of the line contact is therefore 19 mm. Its width is $2a$, the Hertzian contact width, given by Eq. (10).

Normal loads are applied through two adjustable helical springs that are positioned symmetrically on the top of the rider. Time-varying velocity commands are applied to the rotational disk through a servomotor independently, by a function generator. The rider insert and the rotating disk are both made of 52100 bearing steel. The insert was heat treated and water quenched to a hardness of 64 Rockwell C. The rotating disk was left in a spheroidized annealed condition with hardness of 120 Rockwell B. Two combined surface roughnesses

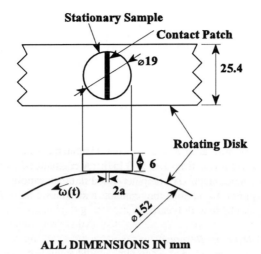

ALL DIMENSIONS IN mm

Fig. 2 Contact details.

having RMS values of 5.0 and 7.1 x 10^{-7} m, were also used. The surface roughness measurements were performed using a profilometer. Sliding velocities ranged from -1 m/s up to 1 m/s. The lubricants used were pure naphthenic mineral oils. Their dynamic viscosities and pressure-viscosity coefficients at atmospheric pressure are listed in Table 1.

Table 1: Lubricant properties.

Lubricant	Dynamic Viscosity, η at 26° C (Pa-s)	Pressure-Viscosity coef., α, x 10^{-8} (Pa^{-1})
A: Naphthenic 100	0.0254	1.954
B: Naphthenic 360	0.103	2.543
C: Naphthenic 1000	0.333	3.043
D: Naphthenic 2000	0.832	3.429

3.2.   Instrumentation

The three components of the contact force (Tangential, F, Normal, W, and Transverse) were measured directly at the contact by a triaxial piezoelectric force transducer. The force measurement bandwidth was from DC to 3 kHz. The transverse component was always negligible.

The disk velocity, V, was measured with a permanent magnet tachometer. Also, the input speed command, V_i, was measured. A linear variable differential transformer (LVDT) was attached to the rider insert, in some tests, to measure tangential motions of the rider directly at the contact.

The electrical contact resistance, R, was measured through a voltage divider network. The measurement bandwidth was from DC to 3 kHz. A constant five volt potential drop was maintained across the contact. A computer-aided data acquisition system performed the manipulations required to obtain the contact resistance. A full description of the instrumentation is available in Polycarpou[27].

3.3.   Apparatus Dynamic Response

Although we have not attempted a complete modal analysis of our system, a few sets of impact tests were performed to identify system natural frequencies and the principal motions of the rider and disk associated with the various vibrational modes. Some understanding of the vibrational characteristics of the system is necessary to appreciate the origins of the oscillations that are observed during the friction experiments.

For illustration purposes, we show, in Fig. 3(a), a frequency response function of the apparatus near the contact in the absence of sliding, i.e., during sticking. Six or eight peaks can be observed. When sliding at low speed, a similar test yields the frequency response function of Fig. 3(b). The response during sliding is nearly identical, although some clear differences can be seen in the frequency range between 900 and 1200 Hz. Evidently, the rider is more weakly constrained during sliding and one of the natural frequencies is significantly reduced. Some type of switching of the system dynamic response must take place between sticking and slipping.

The rider and disk can be represented by a six degree-of-freedom planar model as shown in Fig. 4. X, Y and θ coordinates are the horizontal, vertical and rotational degrees of freedom of the rider. X', Y' and θ' are the horizontal, vertical and rotational degrees of freedom of the disk/shaft assembly. The rider is supported by vertical and horizontal spring elements which are essentially linear. The disk is supported by tapered roller bearings which are the primary compliances in the lower part of the system. These compliances are nonlinear. The two rigid bodies are coupled through a nonlinear (Hertzian) normal contact stiffness. During sticking the contact region also has a finite tangential stiffness. Also, there is a kinematic constraint associated with the circular profile of the disk which the rider must follow as it moves. This constraint requires that small motions of the rider in the X direction are accompanied by (also small) rotations of the rider as it follows the contour of the disk.

In principle, this planar model leads to six vibrational modes, close to the number observed in Figs. 3(a) and 3(b). In addition, elastic modes of the rider, the disk/shaft system, and remaining elements of the frame of the apparatus are present but do not seem to be significant

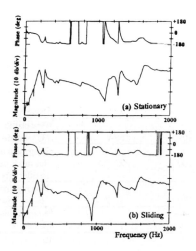

Fig. 3 Frequency response function, lubricant C: (a) stationary, (b) sliding.

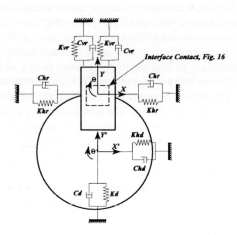

Fig. 4 Disk and rider dynamic model.

Table 2: Experimental system natural frequencies, lubricant A.

MODE	FREQUENCY (Hz)		PRIMARY MOTIONS	COMMENTS
	Stationary	Sliding		
1.	200	200	X, θ, X'	During sliding X and θ follow X'. During sticking X follows X' and angular vibrations θ are observed.
2.	280	280	X, θ, Y'	Two modes. Vertical vibrations of the disk and the drive shaft on their roller bearings. Rocking/ Translational mode whose natural frequency increases with load. Y and Y' are in phase.
3.	325	325	X'	Horizontal vibrations of the disk and the drive the drive shaft on their roller bearings.
4.	800	780	N/A	First elastic mode of the rider associated with the "U" shape of the rider. Usually negligible.
5.	1200	1100	X	Tangential motion of the rider. The stiffness of this mode decreases when sliding, perhaps due to rider offset with respect to center of wheel.
6.	1300	1280	X, θ	Non rigid body mode. Mostly angular vibrations (θ) are observed.
7.	1620	1590	Y	Normal contact resonance. Y and Y' are out of phase.

in the frequency range of interest. Table 2 summarizes the experimental system natural frequencies, during sticking and sliding, at an average applied normal load of 250 N. No clear differences in frequency response were found between the lighter and heavier lubricant. However, after a long dwell period, e.g., overnight, which is long enough for most of the lubricant to be squeezed out from the contact, some changes were observed. Due to the nonlinear character of some of the contact compliances in the system, the linearized dynamic characteristics will be somewhat load-dependent. The mode at 280 Hz is primarily associated with the (uncoupled) vertical vibrations, Y', of the disk against the supporting roller bearings. It is more significant during sticking and less significant during sliding. Another mode at 200 Hz involves a number of degrees of freedom. This mode dominates during sticking, and is also observed during sliding. Between 900 and 1300 Hz we observe mostly tangential (X) and angular (θ) vibrations of the rider. The primary normal contact resonance, a feature of many sliding systems, is at 1620 Hz. Vibrations at or near these frequencies are observed during various phases of the sliding, sticking and during transitions between the two conditions.

Later, in section 6, we will combine a simplified version of the system model, Fig.4, together with a friction model to accurately estimate high frequency friction transients under varying dynamic normal loads.

3.4.   Experimental Results

Unsteady friction experiments under various forced velocity oscillation conditions were conducted. The frequency of oscillation and the DC bias of the input velocity commands were varied, to achieve conditions ranging from quasi-steady, single-valued friction-velocity relations, to unsteady continuous and intermittent sliding, where the tangential force coefficient at a given load and sliding speed can take on a range of values. Two types of intermittent sliding were considered. In the first, periods of sticking are encountered, without motion reversals. The second type is with compete motion reversals. In this case, the transitions from positive to negative velocity and vice-versa can either occur instantly without noticeable sticking, or may involve an intermediate sticking phase, depending on the experimental conditions.

Normal loads during a given test remained essentially constant, as measured by the triaxial force transducer. To investigate the effects of load and lubricant viscosity, four nominal normal loads, ranging from 50 to 500 N, and four lubricants, listed in Table 1, were used. During all experiments the relevant variables of viscosity, normal load, surface roughness, frequency and amplitude of oscillation, were controlled so that the conditions at the contact remained within the boundary and mixed lubrication regimes. Only the lower viscosity lubricants, A and B, will be used for modeling. For the heavier viscosity lubricants C and D, full film conditions were also involved, and thus the friction model, which applies to boundary and mixed lubrication, cannot be used, except at very low velocities.

Fig. 5 shows two sets of typical experimental results under unsteady sliding for lubricant A. The input velocity command to the servodrive is a triangular wave with a frequency of six Hertz, as shown in Fig. 5(a). In the first experiment, shown by the dashed lines in Fig. 5(a), sliding is continuous and unidirectional. In the second experiment, sliding is intermittent with a sticking period, or dwell, of about 45 ms. The normal loads at the contact, remain essentially constant at 250 N, varying slightly (less than 2.5 per cent) due to the out-of-roundness of the disk. The friction forces are shown in Fig. 5(b). When the disk velocity oscillates under conditions of continuous, unidirectional sliding, the friction force is highest at the lowest speeds, with changes in the friction force lagging changes in velocity. This is the delay documented by Hess and Soom[18], and is thought to be due to the fact that the normal separation of the sliding bodies is not able to reach an equilibrium value during most of the velocity oscillation. One observes a squeeze film effect if the sliding velocity changes rapidly. The details of the behavior depend on the dynamics of the system in the direction normal to sliding.

When the macroscopic sliding is interrupted by periods of sticking, as in the second curve of Fig. 5(b), we still note a behavior similar to continuous sliding during most of the slipping phase. However, the transitions to sticking, around $t = 0$ and $t = 0.17$ s are accompanied by significant transient oscillations in the tangential force. These oscillations are predominately at 200 Hz, which corresponds to one of the system resonances, involving tangential and angular motions of the rider. We emphasize that these fluctuations are not really oscillations in the "friction" force, since no gross slip is taking place.

The transition to sticking, which involves a change in tangential contact stiffness from zero to a finite value is quite abrupt. During sticking, the tangential force initially continues to decrease, since the velocity command is trying to drive the disk in the negative direction. However, it is obvious that insufficient torque is being applied to overcome the break-away frictional resistance. Thereafter, the velocity command increases and the tangential force rises

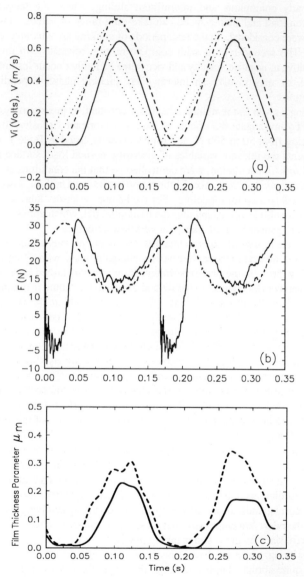

Fig. 5 **Dynamic experimental data, lubricant A, W = 250 N, (- - -) continuous, (——) intermittent sliding:** **(a) velocity input command V_1 (· · ·), sliding velocity V; (b) tangential force F; (c) film thickness parameter** $\hat{h}_o$.

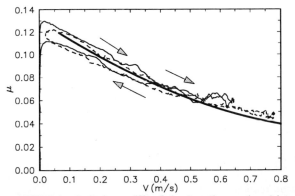

Fig. 6 Experimental friction-velocity loops of Fig. 5: (- - -) continuous, (———) intermittent sliding, (▬▬) quasi-steady sliding.

until, just before reaching its peak value, gross slip begins.

The entire oscillation takes place under boundary or mixed lubrication as confirmed by the contact resistance measurements which remains of the order of ohms, rather than the kilo-ohm range that one would observe with a full fluid film.

Fig. 5(c) shows the film thickness parameter, $\hat{h}_o$, for both experiments. This normal separation measure is calculated from electrical contact resistance measurements, according to Appendix 2 (Eq. A6). At the lower speeds the resistance is small and does not fluctuate very much. At the higher speeds, while remaining low, it increases by one to two orders of magnitude. The fluctuations are relatively large and can be ascribed to some combination of normal contact vibrations and the redistribution of the load at a few different contact points as the asperities on the disk are swept through the contact region. The latter effect is thought to dominate.

Various "friction" force versus velocity relations can be extracted from these tests. The simplest is to plot the tangential force against the disk velocity. Such friction-velocity relations, corresponding to the experiments described above, are shown in Fig. 6. The data from the continuous sliding experiment forms a clockwise loop that reflects the aforementioned delay between changes in velocity and the subsequent changes in the friction force. The equilibrium or steady-state friction-velocity characteristic, based on a curve-fit to data, obtained by changing the velocity very slowly, is shown by the solid line in Fig. 6. The "friction-velocity" relation for the intermittent sliding with stick is also shown. During sticking, the tangential force can take on a range of values.

4.   Friction Modeling

The friction modeling was carried out in two phases. In the first phase[28], single models were developed, where the fluid and solid components of friction do not appear as separate components in the models. These models were then successfully applied to unsteady sliding conditions, emphasizing continuous sliding. As a result of these models, the important variables in characterizing unsteady friction were identified, and also the dimensionless groupings were determined. In the second phase[29], a friction model with separate fluid shear and solid friction components was developed, utilizing the same independent variables and dimensionless groupings as the single models. This model was then used to estimate the instantaneous fluid and solid components of unsteady friction under continuous and intermittent sliding conditions.

4.1.   Single Models

4.1.1.   Model Identification
To establish a relationship between the friction coefficient, μ, the dependent variable, and the independent variables of sliding velocity, V, and normal separation, h_o, a functional identification technique, described in Polycarpou and Soom[28], is used. Procedurally, the functional identification consists of two steps. The first is a correlation test that establishes the functional form of the model. The second step is the parameterization of the model through a nonlinear least square algorithm. This method does not assume a priori knowledge of the model or its functional form.

The empirical models are obtained under steady normal loads and under relatively slowly varying, (less than ten Hertz oscillation), velocities. Under these conditions, it was found[28] (Appendix 2) that the measured contact resistance, R, could be related to the smooth surface EHL theoretical central film thickness, representing an average separation. The model, therefore provides a direct relation between the friction and the normal separation.

In seeking a relationship between $\mu = \mu\,(V,\,h_o)$, the friction coefficient and the sliding velocity as measured from the experiments were used. Since there is a direct relationship between smoothed resistance and the film thickness, the contact resistance was used in determining the model.

4.1.2.   Parameterization
Out of the few hundred candidate functions (models) that were run, two models were selected for parameterization. The first is a "simple" model (in terms of the functional form) and the second is the best performing, so-called, "complex" model. Before proceeding, however, it was necessary to incorporate effects of the various combinations of normal load, viscosity and pressure-viscosity coefficient into the model. The parameterized models which we term the "simple" and "complex" models are
Simple Model:

$$\mu(t) = \frac{c_1' e^{-c_2' V^*(t)}}{1 + c_3' \left(\frac{a}{b}\right)^2 \left(\frac{\eta_i}{\eta_o}\right) R^*(t)} \tag{6}$$

Complex Model:

$$\mu(t) = c_1 + c_2 V^*(t) + c_3 \left(\frac{\eta_i}{\eta_o}\right) \left(\frac{V^*(t)}{R^*(t)}\right)^{0.28} \frac{e^{-c_4 \sqrt{V^*(t)}}}{1 + c_5 \left(\frac{a}{b}\right)^2 \left(\frac{\eta_i}{\eta_o}\right) R^*(t)} \tag{7}$$

Where V^* is a dimensionless velocity parameter given by

$$V^* = \frac{\eta_i \alpha V}{a} \tag{8}$$

R^* is a dimensionless separation parameter, given by

$$R^* = \frac{R}{R_L} \tag{9}$$

$R_L = 35 \ \Omega$, is a limiting contact resistance, determined experimentally. Resistance higher than R_L indicates that operation is beyond the mixed lubrication regime; η_o is a reference viscosity at the inlet conditions; a is the semi-contact width given by Hertzian theory

$$a = \sqrt{\frac{4}{\pi} \frac{W}{b} \frac{r'}{E'}} \tag{10}$$

r' is the effective radius of curvature of the two contacting bodies; E' is the effective elastic modulus. c'_i (i =1-3) and c_i (i =1-5) are empirical constant coefficients fitted to the experimental data and are, for the Simple Model: $c'_1 = 1.25 \times 10^{-1}$; $c'_2 = 2.71 \times 10^5$; $c'_3 = 3.36 \times 10^5$, and for the Complex Model: $c_1 = 7.50 \times 10^{-2}$; $c_2 = -1.20 \times 10^4$; $c_3 = 1.96$; $c_4 = 9.80 \times 10^2$; $c_5 = 9.38 \times 10^3$.

In examining these models, the reader is cautioned against interpreting the coefficient, c_1, as a boundary friction coefficient near zero velocity. We have shown elsewhere[25,26], that when the contact is operating under unsteady sliding conditions, the friction behavior near zero velocity can not be interpreted in this way. This is due to the fact that the separation continuously changes near zero velocity, influencing both the solid and fluid shear components of the friction independently of the sliding speed.

4.2. Two-Component Model

In the previous section (section 4.1.) and in Polycarpou and Soom[28], friction models have been described for a particular line contact geometry operating under a range of conditions. These two-dimensional empirical models describe friction as a function of the sliding velocity and the instantaneous separation of the sliding bodies, normal to the sliding direction. The models were applied to steady and unsteady continuous sliding and to intermittent sliding[30], and estimated the friction behavior quite well.

Although these models captured the essential ingredients of mixed and boundary friction behavior over a wide range of conditions, they have the disadvantage that their functional form is somewhat arbitrary, without a clear physical interpretation. Both the fluid shear and solid components of friction are intertwined in the model, not appearing as separate terms.

The friction model presented in this section[29] employs the same independent variables, with the additional advantage that the fluid shear and solid components of friction are separated. This form of the model provides an unambiguous interpretation of the physical phenomena that take place during unsteady sliding, especially during stick-slip and slip-stick transitions

4.2.1.   Model Identification

To establish a relationship between the friction coefficient, sliding velocity and normal separation, the functional identification technique, described in section 4.1. is used.

Before presenting the mathematical form of the model, it will be helpful to describe in qualitative terms, the different variables that should appear in the two components of friction.

The requirements for the fluid shear component are that it:

(a) be a function of both the instantaneous sliding velocity, the normal separation (film thickness), the contact width and the inlet viscosity, and

(b) diminishes as the sliding velocity goes to zero (i.e., $\mu_f = 0$ at $V = 0$).

A functional form that satisfies the above requirements is the complex term of the previously presented two-dimensional complex model, Eq. (7).

The requirements for the solid friction component are that it:

(a) be a function of the separation only, and not the sliding velocity or the fluid properties and

(b) decrease as the separation increases. When full film is attained the solid friction goes to zero.

To develop the friction model with separate components for fluid shear and solid friction, the following approach was followed: The complex form of the previously developed two-dimensional complex model was adopted as the fluid shear component. Then, using the functional identification technique, various candidate models were examined for the solid friction component, which were added to the fluid shear, to give the total friction. Several experiments under various constant normal loads and different sliding conditions ranging from quasi-steady to unsteady sliding, including sticking and motion reversals, were used.

Among several functions that were evaluated as potential solid friction components of the friction coefficient model, the following function was chosen because of its simplicity and its overall performance

$$\mu_s(t) = \frac{c_6}{1 + c_7 R^*(t)} \tag{11}$$

4.2.2.   Parameterization

To incorporate effects of the additional parameters of normal load, viscosity and pressure-viscosity coefficient into the model, the same dimensionless variables were used as in the single models, Eqs. (6), (7). The parameterized model, which we term the "complex" two-component mixed friction model is

$$\mu(t) = c_a \left(\frac{\eta_i}{\eta_o}\right) \left(\frac{V^*(t)}{R^*(t)}\right)^{0.28} \frac{e^{-c_b \sqrt{V^*(t)}}}{1 + c_c \left(\frac{a}{b}\right)^2 \left(\frac{\eta_i}{\eta_o}\right) R^*(t)} + \frac{c_d}{1 + c_e \left(\frac{a}{b}\right)^2 R^*(t)} \tag{12}$$

where c_i (i = a - e) are empirical constant coefficients fitted to the experimental data. For the conditions considered in this study, these coefficients are: c_a= 2.31; c_b= 1.10 x 10^3; c_c= 3.22 x 10^2; c_d= 7.00 x 10^{-2}; c_e= 3.38 x 10^4. It should be noted that in forming this model we are, in fact, adding solid and fluid friction forces rather than friction coefficients, i.e.,

$$\mu(t) = \frac{F_f + F_s}{W} \tag{13}$$

5.   Simulations and Comparisons With Experiments

5.1. Continuous and Intermittent Sliding

The proposed friction models, Eq. (7) for the single model and Eq. (12) for the two-component mixed friction model, are used to estimate the friction coefficient of the continuous sliding experiment of Fig. 5. Fig. 7 shows the friction coefficient-velocity curves for both models and the experiment as well. Both models capture the multi-friction behavior quite well with the single model being somewhat superior. Similar results are obtained when we apply the models to intermittent sliding. Fig. 8(a) shows the single model estimation on a similar intermittent sliding experiment as the one of Fig. 5, and Fig. 8(b) the two-component model estimating the intermittent sliding experiment of Fig. 5. From these simulations it is evident that, overall, the single model is somewhat superior.

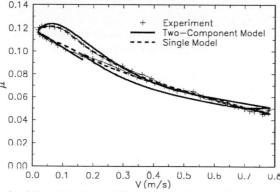

Fig. 7   Friction simulations, lubricant A, W = 250 N, continuous sliding: (+++) experiment, (- - -) single model, (——) two-component model.

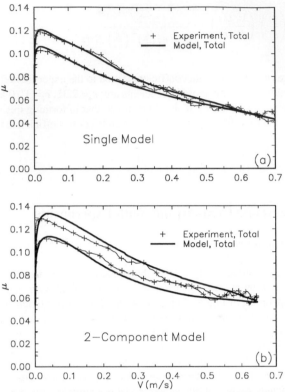

Fig. 8   Friction simulations, lubricant A, W = 250 N, intermittent sliding with stick: (+++)
experiment, (——) model: (a) single, T_d = 15 ms; (b) two-component, T_d = 45 ms (Fig. 5).

As discussed in section 4, the single models have the disadvantage that their functional form
is somewhat arbitrary, without a clear physical interpretation. The two-component mixed
friction model, on the other hand, has the advantage that the fluid shear and solid components
of friction are separated. This form of the model provides an unambiguous interpretation of
the physical phenomena that take place during unsteady sliding.

Next, the two-component mixed friction model with its separate fluid and solid components
is demonstrated, and referred to as the model. Fig. 9(a) shows the estimated and measured
tangential force coefficient for one cycle. Also shown in Fig. 9 are the solid and fluid
components of friction as calculated from the model. When the sliding velocity is near zero
the normal separation is minimum and thus the solid friction is at a maximum. Near zero
velocity, the solid friction is also the total friction coefficient, since there is no fluid shear. As
the sliding velocity increases, the normal separation also increases. As a result, the solid
friction decreases. The fluid shear component, μ_f, is zero during sticking, i.e., zero velocity.

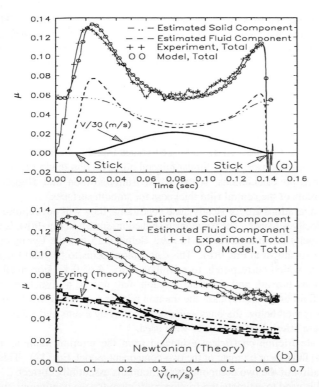

Fig. 9 Friction simulations, lubricant A, W = 250 N, intermittent sliding with stick: (⊤⊤⊤) experiment, (ʊʊʊ) model, (- - -) estimated fluid shear, (- ·· -) estimated solid friction: (a) friction coefficient μ vs. time [includes V]; (b) friction-velocity loops [include traction calculations for Newtonian (×××) and Eyring (ʊʊʊ) fluids].

Exiting the stick regime, as the speed increases, the fluid friction also increases until it reaches a maximum. The fluid friction then decreases, presumably due to thermal (heating) effects, which reduce the viscosity faster than the shear rate increases. During deceleration, the fluid shear increases, until sticking is reached, whereupon it falls to zero again. For this particular combination of load, speed, viscosity and surface roughness, the magnitudes of the solid and fluid friction are nearly equal over the range shown. It should be pointed out that this behavior applies only to the boundary and mixed regimes. Under full film conditions, the solid component will decrease to zero, although this is not apparent in the figure.

Fig. 9(b) shows the friction-velocity loops for the same case. The solid friction loop, obtained by plotting the solid friction (second component of Eq. (12)) versus the sliding velocity, depicts a fairly uniform loop, i.e., with nearly the same breadth at all velocities. The fluid shear-loop, on the other hand, shows a broader loop at lower than higher speeds. The transitions from stick-to-slip are less abrupt in the model than in the experiment. The traction relations given by Evans and Johnson[22] and Johnson and Greenwood[31] are used to predict the fluid friction coefficient for the case just described. Since operation is restricted to the Newtonian and Eyring regimes, the elastic compliance of the metal surfaces is neglected. Therefore, the friction coefficient predicted from these relationships is the steady-state fluid shear component of the total friction. To use the constitutive relationships we need to know the regime in which we are operating and also the fluid properties, including the viscosity, η, and Eyring stress, τ_o. To find η and τ_o the method suggested by Evans and Johnson[21] is followed. That is, η and τ_o are obtained by fitting a relationship between the estimated fluid friction coefficient and the shear strain rate. The values found are comparable with those reported by Evans and Johnson, for a similar mineral oil. The film thickness, which is also needed in the traction calculations, is calculated from Dowson and Higginson[32], and represents an estimate of the central film thickness for smooth surfaces.

Since the traction relations are for steady-state conditions, they will not capture the multi-valued friction-velocity behavior. The Newtonian prediction, shown in Fig. 9(b), is valid from the maximum velocity down to around 0.15 m/s. Below 0.15 m/s the Eyring constitutive relation is used, also shown in Fig. 9(b). This calculation seems to be valid from 0.15 m/s down to 0.02 m/s, which corresponds to a lambda ratio, Λ of 0.1. For small Λ values, asperity interactions prevail, and the fluid traction may not be a function of the bulk properties of the fluid only. In our case the traction relations appear to work satisfactorily at Λ values (≈ 0.1) well below the ones suggested (≈ 0.5) by Evans and Johnson[23]. Also, Hertzian rather than micro-EHL pressures were used.

The steady-state friction coefficient estimated from the traction relations is in close agreement with the fluid shear estimate of the two-component model, Eq. (12). This confirms that our interpretation of the two components of friction is essentially correct.

Next, the model is used to estimate the friction coefficient for the continuous sliding case, also shown in Fig. 5. The friction-velocity loops for this case, are shown in Fig. 10. When the velocity is low, both fluid shear and solid friction are maximum of about the same magnitude. When the velocity increases, both components of friction decrease. Again, the shape of the fluid shear closer resembles the shape of the total friction.

We also consider another case with a higher viscosity lubricant (lubricant B), a longer stick time of 62 ms, and a steady normal load of 170 N. With the higher viscosity lubricant one would expect the fluid shear component to be larger. Fig. 11 shows the experimental, as well as the simulated friction-velocity loops. As expected, the fluid shear is now considerably larger than the solid friction at low velocities, i.e., from zero to about 0.2 m/s. Also note that the solid friction component is smaller than in the previous cases. This is due to the separation being larger.

5.2.   *Complete Motion Reversals*

Next, will apply the model to complete motion reversals. The cases that will be presented here involve symmetric or nearly symmetric motion reversals, which are the most commonly encountered in practice.

Fig. 12 shows an experimental result with a nearly symmetric motion reversal with lubricant

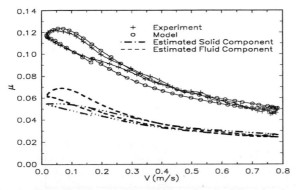

Fig. 10 Friction simulations, lubricant A, W = 250 N, continuous sliding: (₸₸₸) experiment, (ᴏᴏᴏ) model, (- - -) estimated fluid shear, (- ·· -) estimated solid friction.

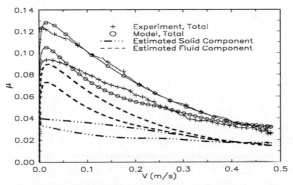

Fig. 11 Friction simulations, lubricant B, W = 170 N, intermittent sliding with stick: (₸₸₸) experiment, (ᴏᴏᴏ) model, (- - -) estimated fluid shear, (- ·· -) estimated solid friction component.

A and a steady load of 270 N. The input velocity command and the sliding velocity for two cycles are shown in Fig. 12(a). The input command is a triangular wave. Motion reversal is abrupt with brief sticking periods of about 1 ms. At the reversal, we observe a brief oscillation with a frequency of 500 Hz, which corresponds to a system resonance.

The friction-velocity loop for this experiment, as well as for two other experiments under similar conditions of speed and lubricant, but different normal loads, are shown in Fig. 13. For the experiment at the 250 N load, also in Fig. 12, the broadest loops are observed, corresponding to the strongest squeeze film effect. For the second experiment, with a load of 170 N, the peak friction is not captured by a friction coefficient, e.g., the typical boundary or static friction model. In both cases, operation is in the boundary and mixed regimes.

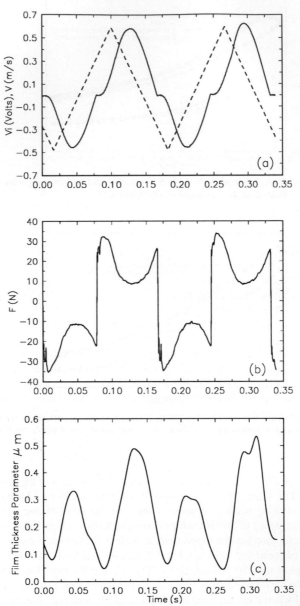

Fig. 12   Dynamic experimental data, lubricant A, W = 250 N, momentary reversal: (a) (- - -) velocity input command, V_i, (——) sliding velocity, V; (b) tangential force, F; (c) film thickness parameter, $\hat{h}_o$.

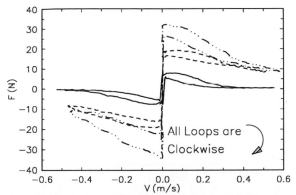

Fig. 13 Experimental momentary reversal friction-velocity loops, lubricant A, normal loads W: (——) 70 N, (- - -) 170 N, (- · · -) 250 N.

When the load is further decreased to 70 N, a full film, as measured by the contact resistance, was achieved at velocities beyond ± 0.4 m/s.

The model is then used to estimate the components of friction for the experiment of Fig. 12. Fig. 14(a) shows the experimental and simulated friction coefficients versus time and Fig. 14(b) shows the friction-velocity loops. The model performs somewhat better in the positive than negative direction. This is ascribed to a small geometric asymmetry in the experimental set-up.

It is also of considerable interest to examine the friction-velocity characteristics of momentary reversals for different lubricants. Fig. 15 shows the experimental friction-velocity loops for 3 similar experiments at a 250 N normal load and lubricants A, C and D. As the viscosity is increased, the friction forces decrease. However, there are also changes in the shapes of the "friction-velocity" characteristics. The velocity at which the maximum friction occurs during acceleration, does not vary monotonically with viscosity. Clearly, the two higher viscosities (C and D) lead to more desirable (positive slope), friction speed characteristics in the 0 to 0.1 m/s sliding speed range. The friction-velocity relation during deceleration remains quite flat for the heavier lubricants. The solid and fluid friction components interact in a complex manner during unsteady sliding in general, and motion reversals in particular.

6.   Unsteady Normal Loads

Friction models are used in the simulation and design of many dynamic and control systems[33,34]. One serious weakness of current models is that they are mostly one dimensional functions of sliding velocity. In section 4[28,29], two-dimensional quasi-steady friction models have been presented that explicitly include the normal separation of the sliding bodies. To

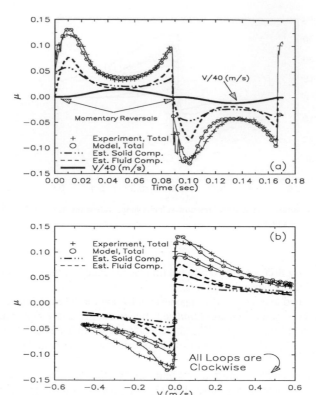

Fig. 14 Friction simulations, lubricant A, W = 250 N, momentary reversal: (+++) experiment, (ooo) model, (- - -) fluid shear component of friction, (- ·· -) solid friction component: (a) friction coefficient μ vs. time [includes V]; (b) friction-velocity loops.

employ these friction models under dynamic loading conditions, the models must be combined with the normal dynamics of the sliding system.

In this section, which is based on Polycarpou and Soom[35,36], the nonlinear normal dynamics of the friction test apparatus are described by a linearized model at a particular steady loading and sliding condition in a mixed or boundary-lubricated regime. The Hertzian bulk contact compliance and film and asperity damping and stiffness characteristics are included as discrete elements. First, a fifth order model is suggested for the normal dynamics of the system. Second, the system model is combined with the simple single friction model, Eq. (6), developed independently, to describe dynamic friction forces under both harmonic and impulsive applied normal loads.

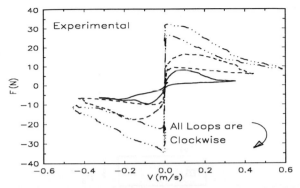

Fig. 15 Experimental momentary reversal friction-velocity loops, W = 250 N, lubricants: (- · · ·) A, (- - -) C, (———) D.

6.1. System Model

In section 3.3. the apparatus dynamic response was presented, and a six degree-of-freedom model was suggested (Fig. 4) to represent the apparatus. To model dynamic friction, we are principally interested in the relative normal motions at the contact. This is in view of the friction models that were presented in section 4. Only the normal degrees of freedom (Y, Y') of the system will be considered here. This is an approximation of the system depicted in Fig. 4, but, as will become clear below, it adequately captures the normal dynamics of the apparatus and the instantaneous separation of the sliding bodies, which are essential for predicting the friction coefficient. Recall that the bulk Hertzian normal contact stiffness, the film characteristics, and the stiffness of the lower part of the system are nonlinear. Only a linearized model will be considered in this work. The linearization is about a steady normal load. While this is also an approximation of the physical system, we will see that the essential mechanics of the problem are captured.

Considering these simplifications, the resulting system model to be considered is shown in Fig. 16. The contact is loaded by the rider weight, by an external mean (steady) load, W, and by a general dynamic load, P_r. P_r can be a harmonic, a band limited white noise, a step change or an impact load. The displacements of the rider y_r, the disk y_d, and the coordinates at the contact, y_1 and y_2, are measured from their equilibrium positions during sliding, at the load and speed at which the system is linearized. The equations of motion during contact, obtained from summing forces on the masses and coordinates, result in a fifth order system model.

The parameters in the equations of motion of the system were estimated using the Eigensystem Realization Algorithm (ERA)[35,37], and also by classical experimental modal analysis techniques. The linearized system model, Fig. 16, was then tested with a very high frequency impact force excitation on the rider. This is done using an impact hammer. Fig. 17(a) shows the impact force applied during an experiment with lubricant B, a mean load of 160 Newtons, and a steady sliding speed of 0.07 m/s. Using the measured force, as the input

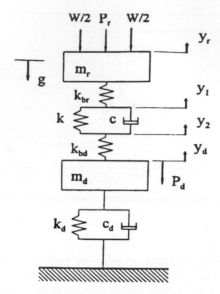

Fig. 16   System dynamic model.

to the model, the predicted acceleration of the rider, is compared against the experimental result, and shown in Fig. 17(b). The model performs reasonably well. The predicted initial accelerations are somewhat larger than the experimental ones.

6.2.   *Friction Transients*

Empirical friction models were obtained under steady normal loads and under relatively slowly varying (less than ten Hertz oscillation) velocities. Under these conditions, it was found that the measured contact resistance could be related to the smooth surface film thickness (Appendix 2), providing, in the model, a direct relation between the friction and the normal separation. In this section, the sliding velocity is held constant while the normal load is changed dynamically, with frequency content beyond 2000 Hz. The simpler model, Eq. (6), will be used here, since both simple and complex models provide equivalent results under the test conditions of the present work.

In the presence of dynamic normal loads, both the film thickness, h_o, and the half-contact width, a, vary with time. The friction model itself however can be considered to be quasi-steady.

The dynamic load, $P_r(t)$ is applied to the rider mass in the system model and the motions y_r, y_1, y_2 and y_d are obtained analytically. The instantaneous film thickness is then calculated from

$$h_o(t) = y_2(t) - y_1(t) + h_{oss} \tag{14}$$

where h_{oss} is the film thickness at the operating point, calculated at the steady state load and speed.

The instantaneous normal load at the contact, $W(t)$, is obtained from the model by computing the force in the spring k_{br}, that is

$$W(t) = k_{br}[y_1(t) - y_r(t)] + W_{ss} \tag{15}$$

where, W_{ss} is the mean load. To further test the accuracy of the dynamic model, the estimated instantaneous normal load will be compared with the measurements.

The instantaneous friction coefficient, $\mu(t)$, is then estimated by substituting for $h_o(t)$ and $a(t)$ in Eqs. (6) and (A5). That is,

$$\mu(t) = \frac{c_1' e^{-c_2' V^*(t)}}{1 + c_3' \left(\frac{a(t)}{b}\right)^2 \left(\frac{\eta_i}{\eta_o}\right) \left(1 - e^{-d \frac{h_o(t)}{\sigma \frac{a(t)}{b} \sqrt{\frac{\eta_i}{\eta_o}}}}\right)} \tag{16}$$

Next, we compare the results of the combined friction and dynamic models with measurements. Consider the same test of Fig. 17 in which a 54 N impact is applied to the rider. The dynamic contact force is shown in Fig. 17(c), along with the force that is estimated using the normal dynamic model. The initial force peak is modeled precisely and the subsequent force oscillations are modeled with a small error in the 2 to 10 ms time interval. The corresponding measured and modeled friction coefficients are shown in Fig. 17(d). Initially, as the load increases, the friction coefficient decreases more than 20 percent. Subsequently, the measured friction coefficient oscillates with a somewhat smaller amplitude than the modeled value.

When the system is excited at other frequencies, away from system resonances, similarly good results are obtained. If the rider is excited at a frequency near one of the system resonances, the behavior becomes more complex. System nonlinearities and coupling among motions can result in nonlinear resonances and larger amplitude dynamic motions that are difficult to describe. Although the friction model will generally remain valid under such conditions, the complexity of the dynamics of the entire system is likely to limit applicability of our modelling efforts.

It should be noted that Eq. (14) is counter-intuitive, since it basically says that the instantaneous normal approach of the sliding bodies is accompanied by a corresponding increase in the film thickness. The physical explanation of this increase in the average film thickness, is that a dimple is formed at the center of the contact, entrapping fluid.

There remains the question: Under what conditions is Eq. (14) valid? For low frequency normal excitations, (below 10 Hz in our apparatus), the instantaneous change in separation is simply equal to the approach (section 5). The oscillations are slow enough that fluid flows out of the contact region as the normal load increases. Under such conditions, the phase difference between the time-varying load and the approach is close to zero. Also, the phase difference between the normal input force, W, and the friction coefficient, μ, is close to zero.

As the frequency of normal load fluctuations is increased, an increasing phase shift between W and μ, is observed. The friction force lags the input load due to the squeeze film effect. Similarly the friction coefficient, which is simply the friction force divided by the total normal load ($W + P_r$), lags both the input and the friction force.

The case presented in this section is a high frequency region where the phase difference between W and μ has reached 180°, i.e., they are out of phase. In this case the normal approach is $[y_1(t) - y_2(t)]$, but the instantaneous film thickness, $h_o(t)$, is calculated using Eq. (14).

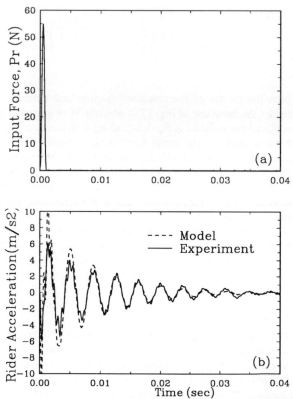

Fig. 17   Dynamic normal excitation: impact force of 54 N, lubricant B, W_m = 160 N, V_m = 0.07 m/s:
(a) input Force; (———) experiment, (- - -) model: (b) rider acceleration; (c) contact force; (d) friction coefficient.

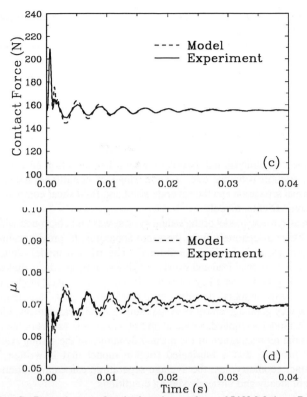

Fig. 17 (continued) **Dynamic normal excitation: impact force of 54 N, lubricant B, W_{st} = 160 N, V_{st} = 0.07 m/s: (a) input Force; (——) experiment, (- - -) model: (b) rider acceleration; (c) contact force; (d) friction coefficient.**

In an intermediate frequency range, or transition range, between these two extreme cases, one finds a phase difference less than 180° and the relationship between approach and instantaneous normal separation is more complicated. The transition range for a given geometry depends on the operating parameters, such as normal load, lubricant viscosity, and sliding speed.

7. Conclusions

Both single and two-component, two-dimensional friction models at a lubricated line contact, operating in boundary and mixed lubrication regimes, have been presented. The friction coefficient is made-up of the solid and the fluid shear components. The solid component is due to the asperity interactions and the fluid friction from the shearing action of the lubricant present at the interface.

The models have been applied to quasi-steady sliding and unsteady continuous and intermittent sliding, including sticking and momentary reversals of motion. The models describe friction behavior under all these conditions quite well. With the two component model it becomes possible to track the instantaneous fluid shear and solid friction components. The fluid shear component is also compared with steady-state predictions using the constitutive relations of other researchers for traction calculations in EHL contacts. Both the solid and fluid friction components contribute to the multi-valued friction behavior during unsteady sliding.

The ability of the two-dimensional, two-component mixed friction model to accurately estimate dynamic friction behavior over a wide range of conditions, indicates that the inclusion of the normal separation and the separate solid and fluid shear components, are key elements of a dynamic friction model.

A linearized normal dynamic model of the sliding system was and combined with one of the friction models, obtained separately from the same apparatus, to provide estimates of the friction under highly dynamic loading conditions. The friction model, developed from measurements at constant normal load and slowly varying sliding speeds, includes the normal approach of the sliding bodies as a key variable. From the combined model, estimates of friction under transients associated with short duration impacts are compared with measurements. The very good agreement between estimates and measurements indicates that, while a good friction model is required, a crucial and most variable aspect of the modeling of dynamic friction is the representation of the normal dynamics of the sliding system.

We therefore propose that a lubricated friction model that is written in terms of instantaneous contact geometry, i.e., the average separation, can capture friction behavior over a wide range of steady and unsteady sliding conditions.

8.   References

1.   F. P. Bowden and D. Tabor, *Friction and Lubrication* (John Wiley, New York, 1956).
2.   D. Tabor, *ASME Journal of Lubrication Technology* **34** (1981) 169.
3.   D. P. Hess and A. Soom, *ASME Journal of Tribology*, **114(3)** (1992) 567.
4.   D. P. Hess and A. Soom, *Journal of Sound and Vibration*, **153(3)** (1992) 491.
5.   D. P. Hess and A. Soom, in *Fundamentals of Friction*, ed. I. Singer and H. Pollock (Kluwer Academic Press, 1992).
6.   J. T. Oden and J. A. C. Martins, *Computer Methods in Applied Mechanics and Engineering,* **52** (1985) 527.
7.   J. A. C. Martins, J. T. Oden and F. M. F. Simoes, *International Journal of Engineering Science*, **28** (1990) 29.
8.   B. J. Hamrock, *Fundamentals of Fluid Film Lubrication* (McGraw-Hill, New York,

1994).

9. R. Stribeck, *Zeitschrift des Vereines Seutscher Ingenieure,* **46(38)** (1902) 1342.

10. S. A. McKee and T. R. McKee, *SAE Journal,* **31** (1932) 371.

11. D. J. Schipper, P. H. Vroegop and A. W. J. de Gee, in *5th International Congress of Tribology EUROTRIB 89* (Helsinki, Finland, **2**, 1989) p. 171.

12. D. J. Schipper and A. W. J. de Gee, *ASME Journal of Tribology,* **117(2)** 250.

13. K. L. Johnson, J. A. Greenwood and S. Y. Poon, *Wear* **19** (1972) 91.

14. J. A. Greenwood and J. B. P. Williamson, *Proceedings of the Royal Society of London* **A295** (1966) 300.

15. S. M. Rohde, in *Surface Roughness Effects in Hydrodynamic and Mixed Lubrication,* (ASME-The Lubrication Division, 1980), p. 19.

16. N. Patir and H. S. Cheng, *ASME Journal of Lubrication Technology,* **100** (1978) 12.

17. J. A. Greenwood and J. H. Tripp, *Proc. Instn. Mech. Engrs.,* **185(1)** (1971) 625.

18. D. P. Hess and A. Soom, *ASME Journal of Tribology,* **112(1)** (1990) 147.

19. A. Harnoy, B. Friedland and H. Rachoor, *Wear* **172** (1994) 155.

20. K. L. Johnson and J. L. Tevaarwerk, *Proceedings of the Royal Society of London,* **A356** (1977) 215.

21. C. R. Evans and K. L. Johnson, *Proc. Instn. of Mechanical Engineers,* **C200(C5)** (1986) 303.

22. C. R. Evans and K. L. Johnson, *Proc. Instn. of Mechanical Engineers,* **C200(C5)** (1986) 313.

23. C. R. Evans and K. L. Johnson, *Proc. Instn. of Mechanical Engineers,* **C201(C2)** (1987) 145.

24. S. Bair and W. O. Winer, *ASME Journal of Lubrication Technology,* **101** (1979) 258.

25. A. A. Polycarpou and A. Soom, in *Friction-Induced Vibration, Chatter, Squeal, and Chaos,* ed. R. A. Ibrahim and A. Soom (ASME Winter Annual Meeting, Anaheim, DE **49**, ASME New York, 1992), p. 139.

26. A. A. Polycarpou and A. Soom, *ASME Journal of Vibrations and Acoustics* **in press**

27. A. A. Polycarpou, *Transitions Between Sticking and Slipping at Lubricated Line Contacts* (MS Thesis, State University of New York at Buffalo, 1992).

28. A. A. Polycarpou and A. Soom, *ASME Journal of Tribology* **117(1)** 178.

29. A. A. Polycarpou and A. Soom, *ASME Journal of Tribology* **in press**.

30. A. A. Polycarpou and A. Soom, *Wear* **181-183(I)** (1995) 32.

31. K. L. Johnson and J. A. Greenwood, *Wear* **61** (1980) 353.

32. D. Dowson and G. R. Higginson, *Elastohydrodynamic lubrication,* 2nd edition, (Pergamon Press, Oxford, 1977).

33. B. Armstrong-Hélouvry, *Control of Machines with Friction* (Kluwer Academic Publishers, Boston, 1991).

34. B. Armstrong-Hélouvry, P. Dupont and C. de Wit Canudas, *Automatica* **30(7)**

(1994) 1083.

35.   A. A. Polycarpou and A. Soom, *ASME Journal of Tribology* **117(2)** (1995) 255.
36.   A. A. Polycarpou and A. Soom, *ASME Journal of Tribology* **117(2)** (1995) 261.
37.   A. A. Polycarpou, *Two-Dimensional Modeling of Boundary and Mixed Dynamic Friction at Lubricated Line Contacts*, Ph.D. Dissertation, State University of New York at Buffalo, 1994.
38.   M. J. Furey, *ASLE Transactions* **4** (1961) 1.
39.   S. Bair and W. O. Winer, *ASME Journal of Lubrication Technology* **104** (1982) 382.

Appendix 1: Nomenclature

A_f	Fluid area of the contact, m^2
A_r	Real area of contact, m^2
a	Hertzian semi-contact width $[a = \{(4Wr')/(\pi bE')\}^{1/2}]$, m
b	Length of the contact, m
c_i $(i = 1 - 5)$	Empirical constant coefficients of the complex, single friction model, Eq. (7)
c_i $(i = a - e)$	Empirical constant coefficients of the two-component friction model, Eq. (12)
c_i' $(i = 1 - 5)$	Empirical constant coefficients of the simple, single friction model, Eq. (6)
d	Empirical constant coefficient of the R, h_o relationship (Eq. (A5))
dV/dh_o	Shear rate, s^{-1}
E	Young's modulus of elasticity, Pa
E'	Effective elastic modulus $[1/E' = (1 - \nu_1^2)E_1 + (1 - \nu_2^2)/E_2]$, Pa
F	Tangential (Friction) Force, N
F_f	Fluid friction force, N
F_s	Solid friction force, N
g	Acceleration due to gravity, m/s^2
h_o	Theoretical central (average) film thickness, Eq. (A1); Instantaneous film thickness (normal separation) $[h_o = y_2 - y_1 + h_{oss}]$, m
$\hat{h}_o$	Film thickness parameter, Eq. (A6), m
h_{oss}	Steady-state film thickness, m
h^*	Dimensionless film thickness, Eq. (A4)
k, c	Stiffness and damping coefficients of the contact, respectively, N/m, N-s/m

k_{bf}, k_{bd}	Bulk stiffness coefficients of the rider and disk, respectively, N/m
k_d, c_d	Stiffness and damping coefficients of the disk, respectively, N/m, N-s/m
L	Lubrication number $[L = \eta V/(pRa_t) = H/Ra_t]$; H Operational parameter, m
m_r, m_d	Mass of rider and disk, respectively, Kg
P_r	Dynamic normal load applied on the rider, N
P_d	Dynamic normal load applied on the disk, N
p	Mean Hertzian contact pressure $[p = W/(2ab)]$, Pa
R	Electrical Contact Resistance, Ω
R^*	Dimensionless separation parameter $[R^* = R/R_L]$
R_L	Limiting contact resistance, Ω
Ra_t	Combined CLA surface roughness, m
r'	Relative curvature of the two contacting bodies, 1, 2 $[1, r' = 1/r_1 + 1/r_2]$, m
T	Temperature at the contact, $^{\circ}$C
T_d	Dwell (stick) time, s
X, Y, θ	Horizontal, Vertical and rotational coordinates of the rider
X', Y', θ'	Horizontal, Vertical and rotational coordinates of the disk/shaft assembly
y_r, y_d, y_t, y_2	Displacements of the rider, disk, and coordinates 1 and 2, respectively (Fig. 16), m
V	Sliding velocity, m/s
V_i	Input velocity command, Volts
V_{ss}	Steady sliding velocity, m/s
V^*	Dimensionless velocity parameter $[V^* = \eta_i \alpha V/a]$
W	Normal load, N
W_{ss}	Mean normal load, N
α	Pressure-viscosity coefficient, Pa-1
ΔV	Rolling speed $[\Delta V = (V_1 + V_2)/2]$. In this work $[\Delta V = V/2]$, m/s
η	Mean lubricant viscosity at the contact, Pa-s
η_i	Lubricant dynamic viscosity at the inlet conditions, Pa-s
η_o	Reference dynamic viscosity at the inlet conditions, Pa-s
Λ	Lambda ratio (film parameter) $[\Lambda = h_o/\sigma]$
μ	Friction coefficient
μ_f	Fluid shear component
μ_s	Solid friction component
ν	Poisson's ratio
σ	Combined RMS surface roughness $[\sigma = \{\sigma_1^2 + \sigma_1^2\}^{1/2}]$, m
τ_f	Fluid shear stress, Pa
τ_o	Reference (Eyring) stress, Pa
τ_s	Shear strength of the interface material, Pa

Appendix 2: Contact Resistance - Theoretical Central Film Thickness Relation

The relevant normal motions at the interface were quite small, usually ranging from 0.03 to 0.5 μm, and were detected indirectly by contact resistance measurements. The two-dimensional models of boundary and mixed friction presented in this work include two important independent variables. One is the normal separation of the sliding bodies, perpendicular to the sliding direction and the other is the sliding velocity. Electrical contact resistance measurements are used to represent the normal separation of the sliding bodies. Resistance measurements in the boundary and mixed lubrication regimes can only be used to represent the separation of the sliding bodies under "quasi-dynamic" conditions. That is, for resistance fluctuations of few Hertz to few tens of Hertz. High frequency resistance fluctuations do not necessarily represent high frequency relative motion of the contacting bodies. They often represent single asperity interactions that are unrelated to the normal separation of the sliding bodies. In section $6^{35,36}$, a system dynamic model of the sliding bodies was presented and the motions from this model are used directly for the friction estimation.

However, the contact resistance is initially used in the friction models. We need to establish a relationship between the contact resistance and the normal separation. A good measure of the normal separation for lubricated contacts is the central film thickness. To establish a relationship between contact resistance and the theoretical central film thickness additional experiments were performed. These experiments were carried out under steady-state conditions: The normal load and the sliding velocity were nominally constant and the contact resistance was measured. The experiments were repeated for the full range of normal loads, lubricants and sliding velocities of interest in this work.

The general trends, such as load and velocity dependence of the contact resistance agree with the work of other researchers[38]. To quantify the contact resistance-film thickness relationship in the boundary and mixed lubrication regimes, the central film thickness, h_o, for a line contact between smooth surfaces was calculated from the Dowson and Higginson[32] formula. This formula applies to a range of combined rolling and sliding conditions, whereas the present work deals with pure sliding. However, the work reported here strongly suggests that this formula is sufficiently accurate for the purpose at hand. Therefore

$$h_o = 2.5 \, r' \left(\frac{\alpha \eta_i \Delta V}{r'} \right)^{0.70} (2\alpha E')^{0.10} (\alpha p)^{-0.26} \qquad (A1)$$

Where ΔV is the rolling speed given by

$$\Delta V = \frac{V_1 + V_2}{2} \qquad (A2)$$

In this work $\Delta V = V/2$ since the rider is stationary. p is the mean contact pressure and is given by

$$p = \frac{W}{2ab} \qquad (A3)$$

To represent the lubricants, sliding speeds, normal loads and surface roughness under consideration, with a single resistance-film thickness relationship the following dimensionless film thickness quantity, h^* is defined

$$h^* = \frac{h_o}{\sigma \frac{a}{b} \sqrt{\frac{\eta_i}{\eta_o}}} \qquad (A4)$$

Fig. A.2.1 shows the contact resistance, R, versus the dimensionless film thickness, h^*. The data represents two lubricants, (lubricants A and B), four normal loads for each lubricant and a number of sliding velocities. The combined surface roughness for these experiments was about 0.05 μm.. Although there is some scatter in the data, the dimensionless film thickness collapses the data well. The scatter in the data is probably due to the inability of the dimensionless film thickness to accurately account for the different conditions considered, rather than inconsistencies in the experiments.

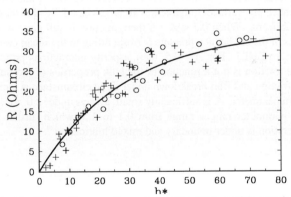

Fig. A.1. Contact Resistance, R, versus dimensionless film thickness, h*: experiments: (+ +) lubricant A; (o o) lubricant B; (——) model.

For the conditions under consideration, the following dimensionless mathematical model describes the relationship between the contact resistance and the film thickness with sufficient accuracy

$$R^* = \frac{R}{R_L} = 1 - e^{-d\frac{h_o}{\sigma\frac{a}{b}\sqrt{\frac{\eta_i}{\eta_o}}}} \qquad (A5)$$

Resistance higher than R_L indicates that we are no longer operating in the boundary and mixed lubrication regimes, or that the model is no longer valid. At higher values the resistance fluctuated rapidly, indicating that full-films were being achieved intermittently. These regimes are considered to be outside the scope of the preset work In our experiments R_L was found to be about 35 Ohms; $d = 0.0245$ is an empirical constant fitted to our data.

When the contact resistance measurement, R, is used to calculate the normal separation, or film thickness parameter, $\hat{h}_o$, as in Figs. 5(c) and 12(c), then Eq. (A5) is rearranged as follows

$$\hat{h}_o = -\frac{\sigma}{b}\frac{a}{d}\sqrt{\frac{\eta_i}{\eta_o}}\, ln\left(1 - \frac{R}{R_L}\right) \qquad (A6)$$

In developing a model for the resistance film-thickness relationship, the theoretical film thickness for smooth surfaces was applied to the rough surfaces studied in this work. Some of the implications of using smooth surface theory to rough surfaces is discussed by Johnson, et al.[13], and Evans and Johnson[23]. The influence of surface roughness on the behavior of elastohydrodynamically lubricated contacts is usually characterized by the film parameter, Λ. Evans and Johnson found that for $\Lambda > 5$, the traction between two rolling and sliding surfaces is negligibly influenced by surface roughness. It is governed by the bulk rheological properties of the lubricant. When $0.5 < \Lambda < 6$ then, traction is still governed by the bulk rheological properties of the oil, but at a pressure corresponding to the mean contact pressure of the asperities (micro-EHL). For $\Lambda < 0.5$, then, asperity interactions prevail (boundary lubrication), and the traction is not a function of the bulk properties of the fluid only. The role played by solid friction and film breakdown under these circumstances remains unclear.

In this work the film parameter, Λ, is sufficiently small, always under 2.0, for all lubricants. Typically, the film parameter ranges range from 0.1 to 1.5, which further supports the observation that operation is under boundary and mixed lubrication[39].

Subject Index

Author Index

ABOUT THE EDITORS

ARDESHIR GURAN was born in Tehran (Iran). B.Sc. (in Structural Engineering) 1981, M.Eng. (in Civil Engineering) 1983, both from McGill University. M.Sc. (in Mathematics) 1989, Ph.D. (in Systems and Control) 1993 from University of Toronto. Visiting Professor at Technical University of Hamburg (Germany), University of Bordeaux (France), Technical University of Vienna (Austria), Virginia Polytechnic Institute (US). Editor-in-chief of *Stability, Vibration and Control of Systems*. Associate editor of *International Journal of Modeling and Simulation*. Research interests: nonlinear dynamics, structural stability, structronics, acoustics, wave propagation, gyroscopic systems, systems and control theory with applications in vehicle dynamics and machine dynamics.

FRIEDRICH PFEIFFER was born in Wiesbaden (Germany). Dipl.-Ing. (in Mechanical Engineering) 1961, Dr.-Ing. (in Mechanical Engineering) 1965, both from Technical University of Darmstadt. Messerschmidt-Boelkow-Blohm 1966. Head of theoretical and technical mechanics department (MBB space division), 1968–75. Project Manager Anti-Ship Guided Missile (MBB guided missile division), 1975. Technical secretary to Dr. Boelkow 1976–77. General Manager Bayern-Chemie GmbH, 1978–80. Vice President R&D (MBB guided missile division), 1980-82. Since 1982, Professor of Mechanics and head of the Institute of Mechanics at Technical University of Munich. Serves on the advisory board of *Ingenieur Archiv, Nonlinear Dynamics, Machine Vibration, Chaos, European Journal of Solid Mechanics, Autonomous Robotics, Stability, Vibration and Control of Systems*. Research interests: nonlinear systems, dynamics and control of mechanical systems, optimization, transmission and robotic systems.

KARL POPP was born in Regensburg (Germany). Dipl.-Ing. (in Mechanical Engineering)1969, Dr.-Ing. (in Mechanical Engineering) 1972, both Technical University of Munich. Assistant to Professor Kurt Magnus (Technical University of Munich), 1969–76. Habilitation in Mechanics, 1978. Visiting Professor at University of California Berkeley and at Universidade Estadual de Campinas (Brasil). Since 1981 Professor of Mechanics and since 1985 head of the Institute of Mechanics at University of Hannover. Research interests: nonlinear vibrations, dynamics, systems and control theory with applications in vehicle dynamics, machine dynamics and mechatronics.